Cholesterol: A Multifaceted Molecule

CHOLESTEROL: A MULTIFACETED MOLECULE

ShivSanjeevi Sripathi and Prerna Pandey

Delve PUBLISHING

www.delvepublishing.com

Cholesterol: A Multifaceted Molecule

ShivSanjeevi Sripathi and Prerna Pandey

Delve Publishing

224 Shoreacres Road

Burlington, ON L7L 2H2

Canada

www.delvepublishing.com

Email: orders@arclereducation.com

ISBN: 978-1-77469-409-1 (Hardcover)

Delve Publishing publishes wide variety of books and eBooks. For more information about Delve Publishing and its products, visit our website at www.delvepublishing.com.

ABOUT THE AUTHORS

ShivSanjeevi Sripathi completed his Masters in Biotechnology from Mumbai University in 2008. He was awarded for academic excellence in both his Bachelors and Masters for securing second rank in Mumbai University in 2006 and first rank in his college: Kishinchand chellaram College. For his Masters he secured first rank in his college KET's V.G.Vaze College. He qualified CSIR and NET and and TOEFL in September 2008. He then worked on a stem cell project at the Specialized Centre for Cell Based Therapy (SCCT),KEM Hospital at Mumbai on a project entitled, : Isolation & detection of stem cells from Human Umbilical cord/ amniotic membrane" following which he worked at Junior Research Fellow at Microbiology & Cell Biology Department, Indian Institute of Science, Bangalore on cloning of cell wall genes and transcription factors in E.coli & M.smegmatis. As a writer, he has authored and co-authored 35 books on various aspects of biology such as bionics, molecular wires, cloning, hypertension, the epidemics of the 21st century, handling depression, camouflage, hygiene, immunology and many more with international publishers. He loves to read and share on interesting aspects of life sciences in books. In his free time he loves to travel and explore and give talks on spirituality, ancient customs and traditions.

Dr Prerna Pandey, a biotechnologist with several years of wet lab research experience, worked at International Center for Genetic Engineering and Biotechnology, New Delhi. Her field of research in PhD included isolation and molecular characterization of geminiviruses, genome sequencing, gene annotation, and gene silencing using the RNA interference technology. She has also worked at Transasia Biomedicals and Advance Enzyme Technologies as a scientist. Prerna has published papers in peer-reviewed journals, and has submitted a number of annotated Geminiviral genome sequences in the GenBank, including two novel ones. She has also completed her editing and proof-reading courses from Society for the promotion of Editors and Proofreaders, UK and now works as a freelance scientific editor and writer. When Prerna is not busy with her assignments, she loves to explore historical places.

TABLE OF CONTENTS

PREFACE

Whether it is a newspaper article or an internet "trend" or countless advertisements on social media-one molecule that most people know is "cholesterol".

"The past year has been truly extraordinary. Not only has there been astonishing medical and public health interest in cholesterol, but the term cholesterol has been brought by the media into common culture and consciousness in quite surprising ways. How did this happen? And what are its implications for science, for medicine, and for public health and policy?"

- Cholesterol Revisited Molecule, Medicine, and Media DeWitt S. Goodman, George Lyman Duff Memorial Lecture, 1989. Arteriosclerosis 9:430-438, July/August 1989: from the lecture given in November 1988, at the 61st Scientific Sessions of the American Heart Association in Washington, D.C.

The unique chemical structure of cholesterol having hydrophilic, hydrophobic, and rigid domains all in one molecule has accounted for its presence across eukaryotic species. The molecule is vitally involved in controlling membrane fluidity and gene transcription. It is the base for all steroid hormones and vitamin D. As we humans face challenges in catabolizing cholesterol, the presence of mutations or excess dietary consumption of this molecule causes its accumulation. This accumulation dates back to mummies and icemen and of course many of us now. However, when there is a cholesterol deficiency, research has implicated how the distribution of vitamin E and K could be challenged.

Given this background, this book is the first of the series on cholesterol. After a brief introduction of the key aspects of its synthesis and processing, research studies have been presented discussing how cholesterol profiles are distributed across the globe-India, the US, China, Asia Pacific and a recent global analysis that speaks of a "redistribution of the cholesterol pattern" from the west to the east.

The book also discusses the connection between cholesterol levels and COVID-19-the virus that has changed our lives from 2020. Early work on coronaviruses outlined the roles of lipids in facilitating their infection. This became pertinent for lipid rafts for angiotensin-converting enzyme 2 (ACE2) receptors of the S-protein of the SARS-CoV-2 virus. Several research studies have been presented outlining how the lipid profile was linked to disease severity and how COVID-19 infections altered the cholesterol profile in patients both during the infection and over long-term follow-ups. There is also a discussion of how vaccination too altered the cholesterol profile on the same lines as that of an infection. This is then followed by how cholesterol-lowering drugs can emerge as potential therapeutics for COVID-19 infections. This opens an exciting chapter in the history of research.

Another section of the book explores how specific genes and interestingly gene interactions can influence the cholesterol profile in our body. Even the response to statins (molecules used to lower LDL-C) has a genetic angle with certain single nucleotide polymorphisms or SNPs controlling such lowered responses. This becomes relevant today given the quest for "precision medicine". The role of gender in cholesterol profiles also has been covered as outlined by research. What was noticed was apart from increased dyslipidemia in females, the control of the same was lower in this gender as unveiled by a few case reports. This necessitates urgent attention to bridging the gap not only in other spheres of life but also in healthcare access.

This book comes at a time when the epidemics of obesity and other lifestyle diseases are plaguing us. The aim of this book is to open up the mind of the reader that cholesterol biology is not just limited to "good" cholesterol and "bad cholesterol but there is a vast arena out there highlighting the multifaceted roles of this molecule that is the base of steroids.

"The most beautiful experience we can have is the mysterious. It is the fundamental emotion that stands at the cradle of true art and true science."\

— Albert Einstein, The World As I See It

CHAPTER 1 Cholesterol: A Snapshot of the Basics

"Fat—not carbohydrate—is the preferred fuel of human metabolism and has been for all of human evolution. We"

— *David Perlmutter, Grain Brain: The Surprising Truth about Wheat, Carbs, and Sugar--Your Brain's Silent Killers*

Contents

1.1 AN INTERESTING RENDEZVOUS WITH THE PAST

"If you're afraid of butter, use cream."

— *Julia Child*

Whether it is the internet or advertisements or daily conversations, we come across terms like "bad cholesterol" and "good cholesterol". Interestingly, Peters and VanSlyke (1946) in *Quantitative Clinical Chemistry* state "although there can be no doubt that deposits of lipids, especially cholesterol, are consistent and characteristic features [of atherosclerotic lesions] there is no indication that hypercholesterolemia plays more than a contributory role in their production". To quote Frantz Jr and Moore (1969); The sterol hypothesis in atherogenesis), "few controversies have divided the medical community so sharply for such a long time as has the sterol hypothesis. The separation between the two points of view has become so extreme that, on the one hand, there are respected scientists who believe that the evidence is already so convincing that further clinical testing is unnecessary, financially wasteful and actually unethical; and, aligned against them, are equally respected scientists who believe that the total weight of evidence accumulated over the many years is too slight to justify further work along these lines".

Back in 1913, Nikolai N. Anitschow documented vascular lesions in rabbits fed purified cholesterol dissolved in sunflower oil that was not seen in controls. The early lesions were Sudan-positive birefringent droplets or liquid crystals of cholesterol esters. Prolonged feeding of cholesterol resulted in connective tissue deposition and fibrous cap formation. He further hypothesized that "cholesterin" (this was the name used in Europe for cholesterol at that time) was entering the artery wall from the blood (Daniel Steinberg, 2004).

Figure 1 below illustrates microscopic observations documented by Anitschow: ***TOP***: Needle-like cholesterol crystals with calcium granules, fibrous cap and foam cells in the plaque in the aorta of a rabbit fed cholesterol for 124 days and then put back on a chow diet for 101 days (Daniel Steinberg, 2004: From N. Anitschkow, Experimental atherosclerosis in animals, E.V. Cowdry (Ed.), Arteriosclerosis: A Survey of the Problem, Macmillan, New York (1933), pp. 271-322. *Attribution 4.0 International (CC BY 4.0): You are free to: Share — copy and redistribute the material in any medium or format; Adapt — remix, transform, and build upon the material for any purpose, even commercially. This license is acceptable for Free Cultural*

Works. The licensor cannot revoke these freedoms as long as you follow the license terms: You must give appropriate credit, provide a link to the license, and indicate if changes were made. You may do so in any reasonable manner, but not in any way that suggests the licensor endorses you or your use.):

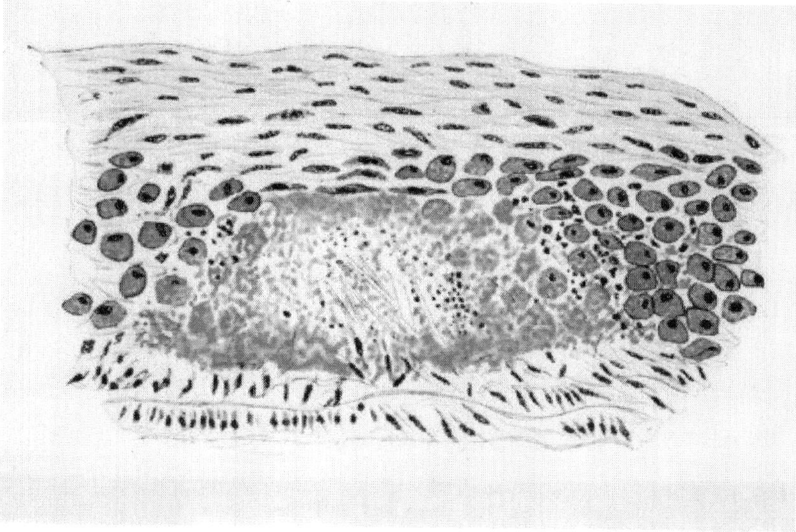

BOTTOM: Upon returning the rabbit to a chow diet for 785 days after 106 days of a cholesterol diet, the partial reversal of the lipids in a plaque was described by Anitschkow describes as: "The lipoidal masses have disappeared; only a few cholesterin (cholesterol) crystals and lipoid-containing wandering-cells are still present." (Daniel Steinberg, 2004: From N. Anitschkow, Experimental atherosclerosis in animals, E.V. Cowdry (Ed.), Arteriosclerosis: A Survey of the Problem, Macmillan, New York (1933), pp. 271-322. *Attribution 4.0 International (CC BY 4.0): You are free to: Share — copy and redistribute the material in any medium or format; Adapt — remix, transform, and build upon the material for any purpose, even commercially. This license is acceptable for Free Cultural Works. The licensor cannot revoke these freedoms as long as you follow the license terms: You must give appropriate credit, provide a link to the license, and indicate if changes were made. You may do so in any reasonable manner, but not in any way that suggests the licensor endorses you or your use.*):

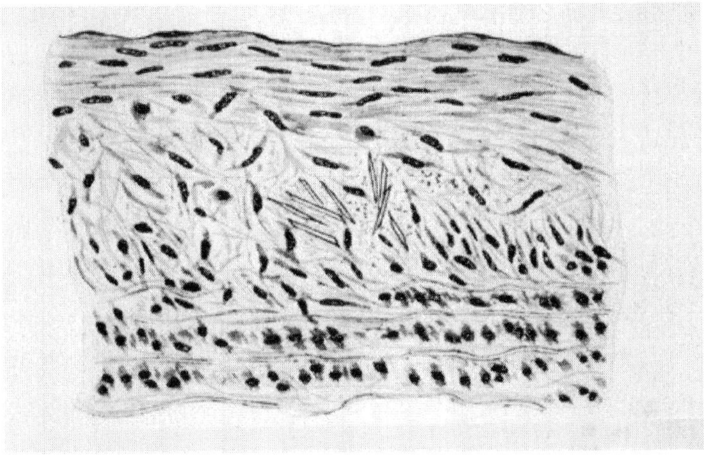

Russ, Eder, and Barr (1951) probed the human serum and reported that α-lipoprotein was elevated in ladies prior to menopause vs. men paving the way for subsequent lipoprotein class discoveries. Subsequently, the randomized trial [the Coronary Primary Prevention Trial (CPPT)] inclusive of "study cohort would consist of ~3,800 men, ages 35–59 years, with no history of coronary heart disease and no signs of current disease. However, these men would be at high risk because of total blood cholesterol levels of 265 mg/dl or higher (men with cholesterol levels in the 95th percentile for this age group)" heralded a new change. Described as a herculean endeavor, the study reported "the number of events (definite coronary heart disease death and/or nonfatal heart attack) was 19% lower in the cholestyramine-treated group, P-value of <0.05". A subsequent Consensus Development Conference on Lowering Blood Cholesterol to Prevent Heart Disease, December 10–12, 1984 at the NIH Clinical Center in Bethesda, Maryland voted "yes" on two questions: "Was blood cholesterol causal? and Would reducing it help prevent heart disease?".

(Steinberg D, 2006)

Figure 2 below illustrates the unanimous conclusion of the 1984 NIH Consensus Panel (Steinberg D, 2006: The pathogenesis of atherosclerosis. An interpretive history of the cholesterol controversy, part IV: the 1984 coronary primary prevention trial ends it—almost: *This is an open access article distributed under the terms of the Creative Commons CC-BY license, which permits unrestricted use, distribution, and reproduction in any medium, provided the original work is properly cited. You are not required to obtain permission to reuse this article*):

NIH CONSENSUS CONFERENCE - PANEL'S CONCLUSIONS

"It has been established beyond a reasonable doubt that lowering definitely elevated blood cholesterol levels (specifically, blood levels of low-density lipoprotein [LDL] cholesterol) will reduce the risk of heart attacks caused by coronary heart disease. . ."

From NIH Consensus Development Conference (JAMA 1985;253:2080)

A brief timeline of the lipid hypothesis that proposed that hypercholesterolemia was a causative factor in human atherosclerosis is presented below in figure 3 (Steinberg D, 2006: An interpretive history of the cholesterol controversy, part V: the discovery of the statins and the end of the controversy: *This is an open access article distributed under the terms of the Creative Commons CC-BY license, which permits unrestricted use, distribution, and reproduction in any medium, provided the original work is properly cited. You are not required to obtain permission to reuse this article*):

1913: Anitschkow, the cholesterol-fed rabbit model
1939: Muäller, familial hypercholesterolemia, xanthomatosis, and angina pectoris linked to hypercholesterolemia
1949: Gofman, the lipoproteins of human plasma and their correlation with the risk of coronary heart disease
1952–4: Kinsell and Ahrens, blood cholesterol in normal subjects is increased by saturated fats in the diet
1955: Keys' Seven Country Study: incidence of coronary heart disease directly correlated to hypercholesterolemia and to dietary fat intake in an international epidemiologic survey
1961: Framingham Study: within a typical American community, coronary heart disease risk is highest in groups with highest blood cholesterol levels
1961: American Heart Association first endorses the "prudent" low-fat diet
1964: Nobel Prize to Konrad Bloch for elucidating the pathway for cholesterol biosynthesis, paving the way for the statins a decade later
1966–9: Leren, Miettinen, Dayton, and others: reducing blood cholesterol levels by reducing saturated fat intake reduces coronary heart disease risk; however, data are not considered persuasive

1974: Goldstein and Brown, the LDL receptor and regulation of cholesterol and lipoprotein metabolism
1976: Endo, discovery of the first effective statin drug (statins not marketed until 1987)
1980: Merck, discovery of mevinolin (lovastatin), later to become the first statin to reach the market
1984: Lipid Research Clinics' Coronary Primary Prevention Trial, which started in 1971, shows significant reduction in coronary heart disease primary events in hypercholesterolemic men treated with cholestyramine
1984: National Institutes of Health (NIH) Consensus Conference on Lowering Blood Cholesterol to Prevent Heart Disease: NIH, for the first time, declares decreasing blood cholesterol to be a national public health goal
1984: NIH coordinates a national program to teach physicians and patients how to diagnose and deal with dyslipidemia, the National Cholesterol Education Program
1985: Nobel Prize to Brown and Goldstein for groundbreaking work on the regulation of cholesterol and LDL metabolism
1994: The statin era: Scandinavian Simvastatin Survival Study, the first truly large-scale, randomized, double-blind trial to show that aggressive treatment (with simvastatin) not only reduces coronary heart disease mortality but also decreases all-cause mortality
1995–present: A stunning series of studies demonstrating the power and the safety of statin intervention

1.2 A QUICK REVIEW OF THE BASICS

Functioning as one of the vital structural components of cell membranes to control their fluidity and serving as a building block for a slew of steroid hormones (cortisol, adrenal androgens and aldosterone), sex hormones (estrogens, progesterone and testosterone), vitamin D, and bile acids, the involvement of cholesterol is being expanded inclusive of cell function regulation. Several proteins also function in the presence of cholesterol like cholesterol-dependent cytolysins that need cholesterol-rich milieus for binding and proteins with sterol binding sites like cleavage-activating protein (SCAP). The unique structure of cholesterol has twenty-seven carbon atoms with a central sterol nucleus of four hydrocarbon rings, a hydrocarbon tail and a hydroxyl group. All steroid hormones have this central ring.

(Benito-Vicente A, Uribe KB, Jebari S, Galicia-Garcia U, Ostolaza H, Martin C, 2018;

Huff, Boyd and Jialal, 2021; Craig, Yarrarapu, Dimri, 2021)

Figure4 below depicts the molecular structure of cholesterol (Huff, Boyd and Jialal, 2021© 2022, StatPearls Publishing LLC. *This book is distributed under the terms of the Creative Commons Attribution 4.0 International*

License (http://creativecommons.org/licenses/by/4.0/), which permits use, duplication, adaptation, distribution, and reproduction in any medium or format, as long as you give appropriate credit to the original author(s) and the source, a link is provided to the Creative Commons license, and any changes made are indicated):

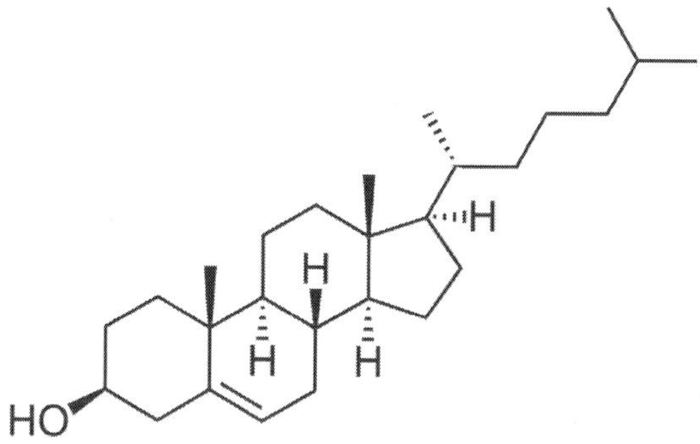

There are two sources of cholesterol in cells: *intracellular synthesized cholesterol* and *dietary cholesterol*.

De novo synthesis is subjected to tight regulation where sterol sensors sterol regulatory element-binding proteins (SREBPs) in the endoplasmic reticulum (ER) are vitally involved. SREBP activation augments the expression of the limiting enzyme HMG-CoA reductase to form cholesterol that is stored as cholesterol esters (CEs) in lipid droplets or distributed. HMG-CoA reductase is subjected to regulation by adenosine monophosphate (AMP)-activated protein kinase (AMPK) and a phosphoprotein phosphatase. The availability of plenty of cholesterol lowers the HMG-CoA reductase activity. When substrates are available (well-fed state), feedback inhibition is suppressed to cause the pathway to get activated. While insulin and thyroxine activate the enzyme, an inhibitory effect is seen with glucagon.

Figure 5 below depicts cholesterol homeostasis (© 2021 Zhou and Sun. *This is an open-access article distributed under the terms of the Creative Commons Attribution License (CC BY). The use, distribution or reproduction in other forums is permitted, provided the original author(s) and the copyright owner(s) are credited and that the original publication in this journal is cited, in accordance with accepted academic practice. No use, distribution or reproduction is permitted which does not comply with these terms*):

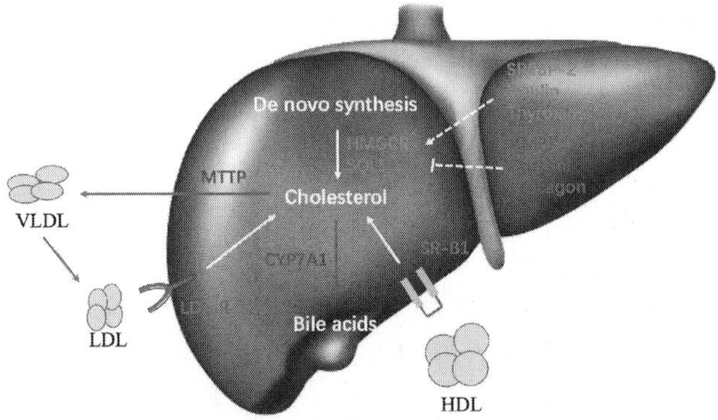

The ability to synthesize cholesterol lies with all cells with the liver emerging as the predominant site of synthesis. Entailing the role of the smooth endoplasmic reticulum (SER), enzymes and proteins, the *de novo* synthesis occurs in the cytoplasm (hepatic). The first step entails the condensation of 2 molecules of acetyl-CoA forming acetoacetyl-CoA. The subsequent action of hydroxymethylglutaryl-CoA synthase (HMG-CoA synthase) adds a third molecule of acetyl-CoA to acetoacetyl-CoA to form 3-hydroxy-3-methyl glutaryl coenzyme A (HMG-CoA). The rate-limiting, committed step follows next where HMG-CoA is reduced to mevalonate by hydroxymethylglutaryl-CoA reductase (HMG-CoA reductase). Subsequent conversions of mevalonate to 3-isopentenyl pyrophosphate, farnesyl pyrophosphate, squalene, and lanosterol ensue. With the involvement of the ER, lanosterol is converted to 7- 7-dehydrocholesterol in several steps and dehydrocholesterolreductaseconverts7-dehydrocholesterol to cholesterol.

Figure 6 below illustrates (A) Subcellular localization of the cholesterol biosynthetic pathway in mammalian cells. (B) SREBP-2 pathway regulation (© 2020 Charles, Shackelford, Faust, Fliesler, Stangl and Kovacs. *This is an open-access article distributed under the terms of the Creative Commons Attribution License (CC BY). The use, distribution or reproduction in other forums is permitted, provided the original author(s) and the copyright owner(s) are credited and that the original publication in this journal is cited, in accordance with accepted academic practice. No use, distribution or reproduction is permitted which does not comply with these terms*):

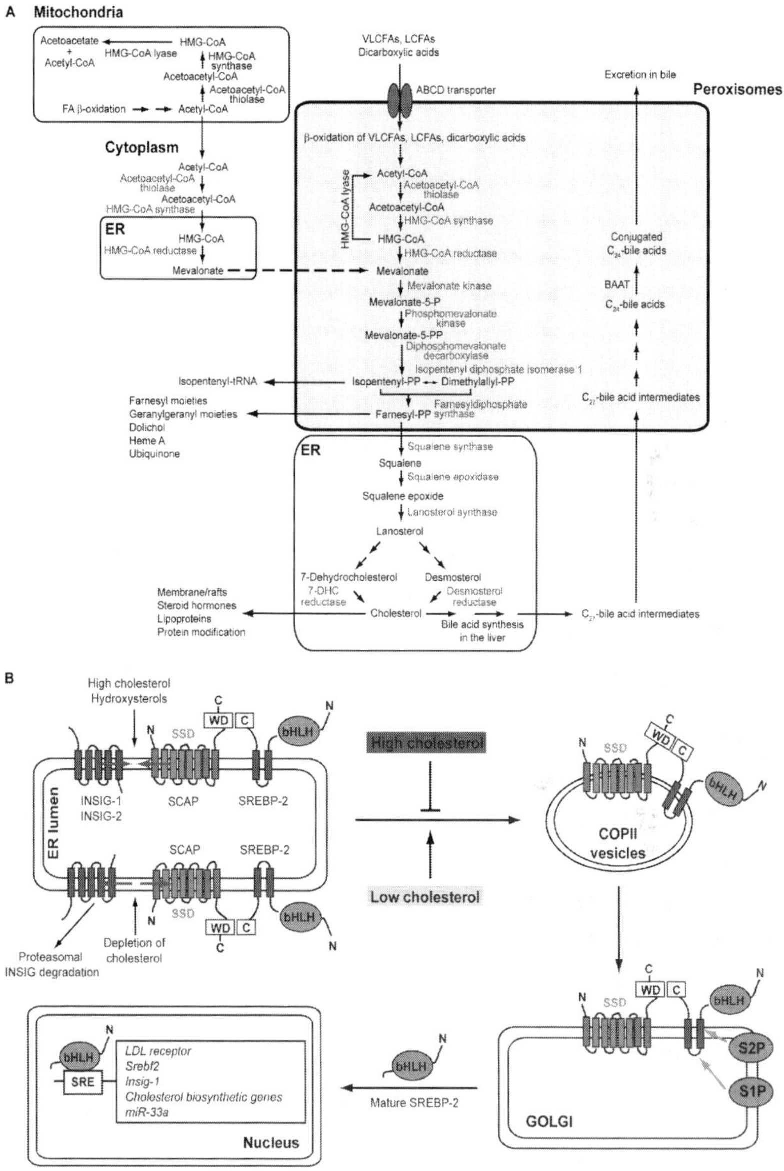

The regulation of HMG-CoA reductase also entails transcriptional control with the transport of the ER-based SREBPs to the Golgi complex. Subsequent processing to an active fragment and binding of the fragment to the SRE (sterol regulatory element) in the nucleus activates HMG-CoA reductase and other enzyme gene expression. In the presence of high cholesterol, another set of ER proteins referred to as 'insigs' (insulin signaling proteins) block SREBP transport to the Golgi complex.

(Afonso MS, Machado RM, Lavrador MS, QuintaoECR, Moore KJ, Lottenberg AM, 2018;

Benito-Vicente A, Uribe KB, Jebari S, Galicia-Garcia U, Ostolaza H, Martin C, 2018;

Huff, Boyd and Jialal, 2021; Craig, Yarrarapu, Dimri, 2021)

The interaction with SCAP and Insig-1 allows the attachment of SREBP-2 (the “master” regulator of cholesterol synthesis and uptake) to the ER membrane. When cholesterol levels dip, proteasomal degradation of Insig-1 ensues to facilitate the sorting of SCAP-SREBP into intoCOPII-coated vesicles via small GTPase, Sar1 and the coat proteins Sec23/Sec24 and Sec13/Sec31. This COPII coated vesicles guide the transport of the complex to the Golgi complex for the release of the NH2-active domain of SREBP by site-1 (S1P) and site-2 (S2P) proteases. The binding of SREBP to SRE results in the synthesis of enzymes and LDLR for uptake (Afonso MS, Machado RM, Lavrador MS, QuintaoECR, Moore KJ, Lottenberg AM, 2018:):

Key: ER: endoplasmic reticulum, SREBP-2: sterol regulatory element-binding protein-2; SCAP: SREBP cleavage-activating protein; Insig-1: Insulin-induced gene protein-1

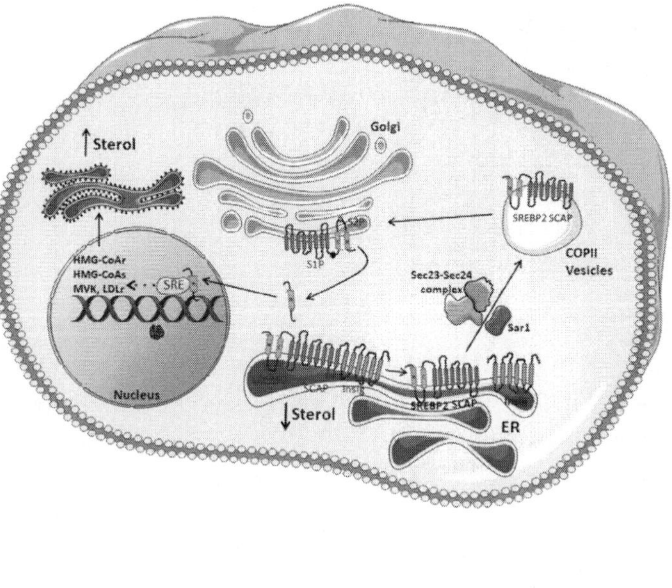

Figure 7 below illustrates the role of insigs in the control of cholesterol metabolism: ***TOP***: A conformational change in SCAP due to binding of cholesterol-SCAP causes the protein to bind insigs to suppress SREBP activation and hence lowered HMG-CoA reductase synthesis. The binding of oxysterols to Insigs suppress SREBP activation and also boost HMG-CoA reductase degradation. Lanosteroldirecty accelerates HMG-CoA reductase degradation (Dong, XY., Tang, SQ. & Chen, JD, 2012: Open Access *This article is published under license to BioMed Central Ltd. This is an Open Access article is distributed under the terms of the Creative Commons Attribution License (https://creativecommons.org/licenses/by/2.0), which permits unrestricted use, distribution, and reproduction in any medium, provided the original work is properly cited*):

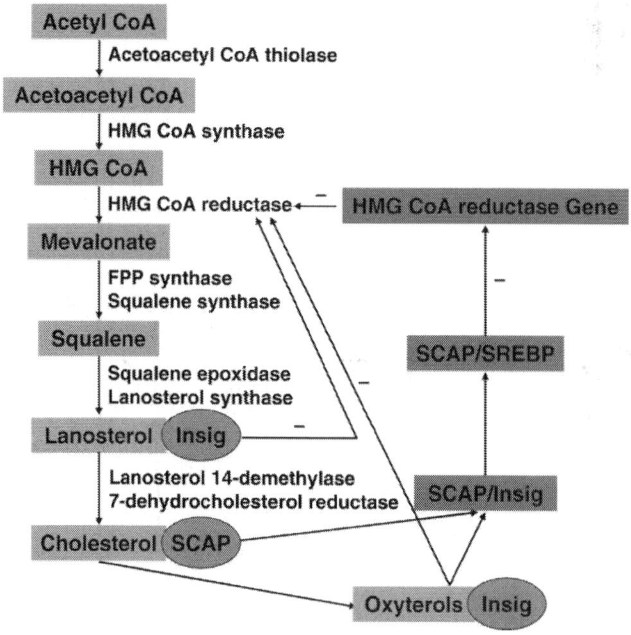

BOTTOM: Sterol overaccumulation causes SCAP-Insig binding to stall SCAP/SREBP complex delivery to the Golgi. The binding of Insigsto HMG-CoA reductase leads to the ubiquitination/degradation of the enzyme to inhibit cholesterol synthesis (Dong, XY., Tang, SQ. &Chen, JD, 2012: Open Access *This article is published under license to BioMed Central Ltd. This is an Open Access article is distributed under the terms of the Creative Commons Attribution License (https://creativecommons.org/licenses/by/2.0), which permits unrestricted use, distribution, and reproduction in any medium, provided the original work is properly cited*):

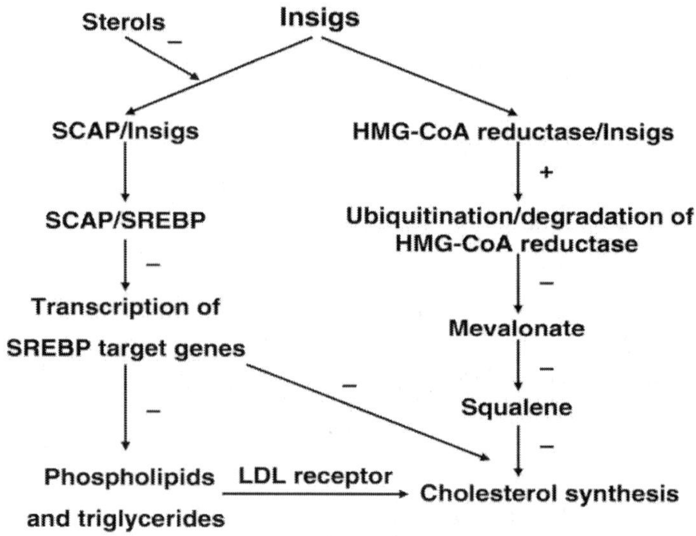

Apart from de novo synthesized cholesterol, the other source of this molecule is **dietary cholesterol absorption**. The diet mainly has lipids such as cholesterol, free fatty acids (FFA) and triglycerides that are absorbed in the intestine. On account of the limited efficiency of this absorption, the cholesterol is emulsified by bile acids to form cholesterol-bile acid micelles for intestinal delivery. Niemann-Pick C1-like 1 (NPC1L1) protein then facilitates free cholesterol uptake by enterocytes once cholesterol esters are hydrolyzed to cholesterol by lipases in the intestine.

Figure 8 below illustrates the cholesterol uptake process (A) Niemann-Pick C1-like 1 in the enterocyte membrane for uptake of cholesterol released from bile acid-cholesterol micelles. This cholesterol reaches the ER and is transported back to the intestinal lumen through transporters like ATP-binding cassette transporters sub-family G member 5 (ABCG5) and member G 8 (ABCG8) with plant sterols or is re-esterified by ACAT or Acyl-CoA acyl-transferase. Subsequent incorporation of cholesterol into chylomicrons and their secretion into the lymph with later loss of triglycerides; they are then taken up by the liver via low-density lipoprotein receptor or LDLR related proteins (B) Conformational change inNPC1L1due to binding with cholesterol for release of the YVNXXF-containing region from the membrane to the cytosol. Clathrin-coated pits facilitate cholesterol internalization of the cholesterol-NPC1L1-Numb complex via a clathrin adaptor protein (Numb) (Benito-Vicente A, Uribe KB, Jebari S, Galicia-Garcia U, Ostolaza H, Martin C, 2018; © 2018 by the authors. Licensee

MDPI, Basel, Switzerland. *This article is an open access article distributed under the terms and conditions of the Creative Commons Attribution (CC BY) license: http://creativecommons.org/licenses/by/4.0/).:*

Key: NPC1L1: Niemann-Pick C1-like 1; ACAT: Acyl-CoA acyl-transferase; Chol ester: Esterified cholesterol; CM: Chylomicrons; LPL: lipoprotein lipases; TG: Triglycerides; FFA: Free fatty acids; LDLR: low density lipoprotein receptor; LRPs: LDLR related proteins

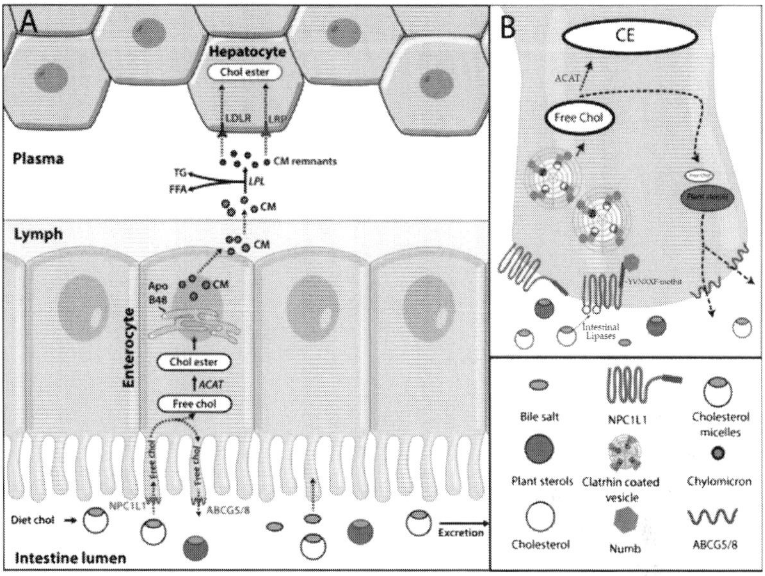

The nuclear receptor Liver X Receptor (LXR) mediates this intestinal cholesterol uptake and secretion pathways. The expression of ABCG5, ABCG8 and NPC1L1 is influenced by the cholesterol sensors LXRα (Nr1h3) and LXRβ (Nr1h2): both are LXR isoforms. The pathways functioning in cholesterol absorption are depicted in figure 9 below with micelle-based transport of free cholesterol (FC). Transfer to the lymph entails chylomicron (CM) formation with cholesteryl ester, triglycerides, and apolipoprotein B-48 (apo B-48)in the Golgi complex (GC). Cholesterol entry back to the lumen entails NPC1L1 inhibition and ABCG5/G8 activation by LXRα and LXRβ (Afonso MS, Machado RM, Lavrador MS, QuintaoECR, Moore KJ, Lottenberg AM, 2018: © 2018 by the authors. Licensee MDPI, Basel, Switzerland. *This article is an open access article distributed under the terms and conditions of the Creative Commons Attribution (CC BY) license: http://creativecommons.org/licenses/by/4.0/*):

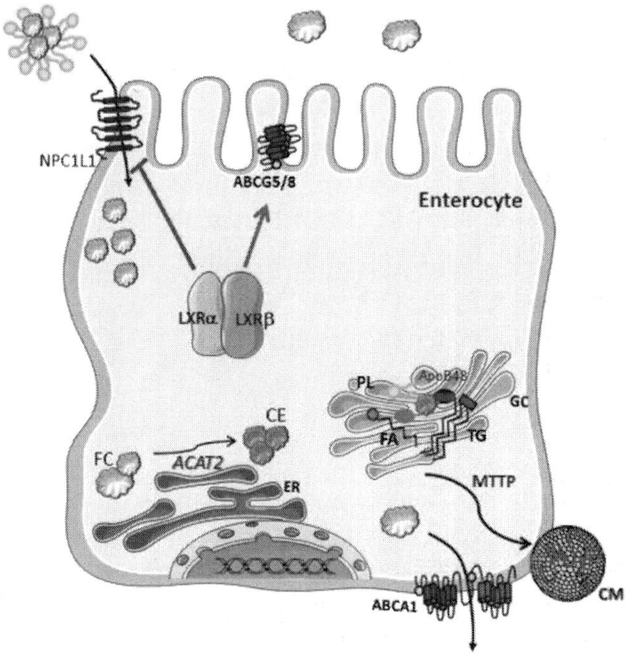

The non-polar nature of the hydrocarbon tail and the central ring of cholesterol prevents its miscibility with water that necessitates the packaging of cholesterol with proteins (apoproteins) forming lipoproteins for transport in the blood. Fasting results in the formation of lipoproteins called chylomicrons containing ApoB48, a truncated form of ApoB100 formed via alternative mRNA editing and other apolipoproteinsApoA-I, ApoA-II, ApoA-IV, ApoA-V, ApoC-I, ApoC-II, ApoC-III or Apo-E, that are incorporated during chylomicron biogenesis or acquired from other circulating lipoproteins. The chylomicrons that are newly synthesized are secreted into the lymph and undergo transport through the lymphatic system to the thoracic duct where draining of the chylomicron-rich lymph into the bloodstream at the left subclavian vein ensues. Interaction of lipoprotein lipases (LPL) of peripheral tissues and these circulating chylomicrons results in triglyceride hydrolysis by LPL activation by ApoC-II in the chylomicrons. Muscle cells and adipocytes then take up the FFAs released. The smaller particles post-FFArelease are cholesterol ester-rich high density lipoproteins (HDL) that get ApoE and transfer ApoA and ApoC to other lipoproteins. ApoEinteractions with low density lipoprotein receptor (LDLR) and other LDLR related proteins (LRPs) results in chylomicron remnant clearance by the liver.

The predominant organ functioning in cholesterol metabolism is the liver apart from synthesizing and clearing lipoproteins. The organ secretes cholesterol in triglyceride-rich lipoproteins called very low density lipoproteins (VLDL). The synthesis of VLDLs entails nascent apoB100translocation across the hepatocyte ER membrane to be lipidated by microsomal triglyceride transfer protein (MTP). Tight control is warranted in the synthesis of these particles for optimal cholesterol distribution. Coat protein complex II (COPII) vesicles facilitate the transport of partially lipidatedVLDL particles to the Golgi where they take up apoA1 and apoEapolipoproteins and additional lipids. The secretion of mature VLDL particles in the bloodstream allows the distribution to other tissues. If the VLDL secretion is in excess, the low density lipoprotein (LDL) levels are boosted that are associated with health risks while lowered VLDL secretion causes accumulation of lipids in the liver to cause fatty liver disease hence, necessitating tight regulation.

In the LDL receptor (LDLR) pathway, intermediate density lipoproteins (IDLs) are formed from VLDL remnants in the plasma that are processed by hepatic lipases to form low density lipoproteins (LDLs) from HDLs via apolipoprotein-lipid exchange. Chiefly constituted by apoB-100 and cholesterol esters, LDLs are the predominant carriers of cholesterol in the system. The binding of LDL-LDLR results in endocytosis of the particles via LDL receptor adaptor protein 1 (LDLRAP1)-mediated NPXY motif (in the LDLR cytoplasmic tail)recognition for LDLR clustering into clathrin-coated pits. Delivery of the LDLR-LDL complex into the endocytic compartment results in dissociation of the LDLR due to the pH change and is recycled to the membrane while the lysosome processes LDL to release free cholesterol that reaches the ER via Niemann-Pick type C1/C2 (NPC1/NPC2) proteins.

(Afonso MS, Machado RM, Lavrador MS, QuintaoECR, Moore KJ, Lottenberg AM, 2018;

Benito-Vicente A, Uribe KB, Jebari S, Galicia-Garcia U, Ostolaza H, Martin C, 2018;

Huff, Boyd and Jialal, 2021; Craig, Yarrarapu, Dimri, 2021)

Figure 10 below ***TOP***: summarizes the scheme of cholesterol transport where VLDLs from the liver are converted to cholesterol-rich LDL particles and are taken up via the LDLR pathway. Surplus cholesterol in peripheral tissues is cleared in HDL particles where ApoA-1 lipid free particles (pre-βHDL) and ATP-binding cassette transporter A1 (ABCA1) interactions facilitate HDL particle maturation that comes back to the liver for clearance

(Benito-Vicente A, Uribe KB, Jebari S, Galicia-Garcia U, Ostolaza H, Martin C, 2018; © 2018 by the authors. Licensee MDPI, Basel, Switzerland. *This article is an open access article distributed under the terms and conditions of the Creative Commons Attribution (CC BY) license: http:// creativecommons.org/licenses/by/4.0/).:*

Key: NPC1: Niemann-Pick C1; NPC2: Niemann-Pick C2; ACAT: Acyl-CoA acyl-transferase; CM: Chylomicrons; LPL: lipoprotein lipases; TG: Triglycerides; LDLR: low density lipoprotein receptor; LRPs: LDLR related proteins; VLDL: very low density lipoproteins; HDL: high density lipoproteins; ABCG1: ATP-binding cassette subfamily G member 1; LCAT: Lecithin-cholesterol acyltransferase; HDL: High density lipoproteins

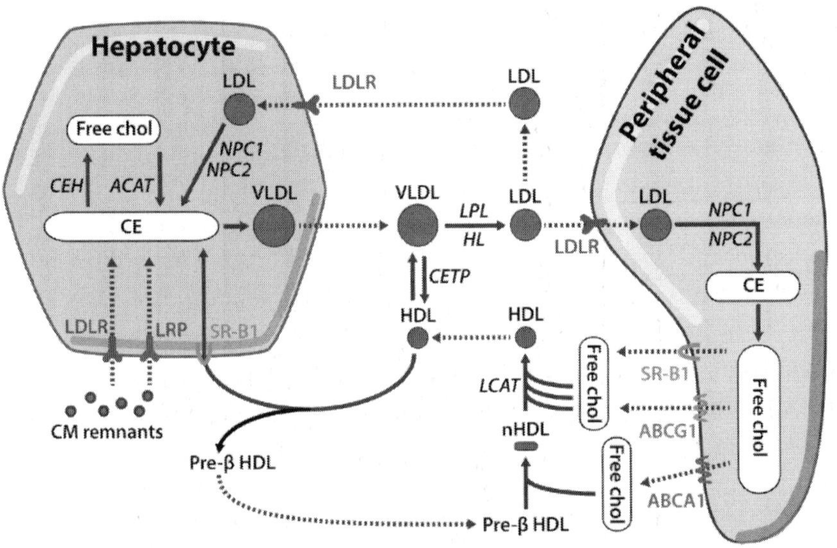

BOTTOM: A brief on LDL processing: (A) The LDLR pathway and mutants (B) Class 2 LDLR mutants where LDLR is retained in the endoplasmic reticulum; (C) Class 3 mutants, where LDL and LDLR do not bind (D) Class 4 mutants where the LDL-LDLR complex internalization is targeted (E) Class 5 mutants that show recycling defects (F) LDL-LDLR binding is targeted due to faulty ApoB-100 (G) proproteinconvertasesubtilisin/kexin 9 (PCSK9: functions in the post-transcriptional control of the LDLR pathway) gain of function mutant(Benito-Vicente A, Uribe KB, Jebari S, Galicia-Garcia U, Ostolaza H, Martin C, 2018; © 2018 by the authors. Licensee MDPI, Basel, Switzerland. *This article is an open access article distributed under the terms and conditions of the Creative Commons Attribution (CC BY) license: http://creativecommons.org/licenses/by/4.0/).:*

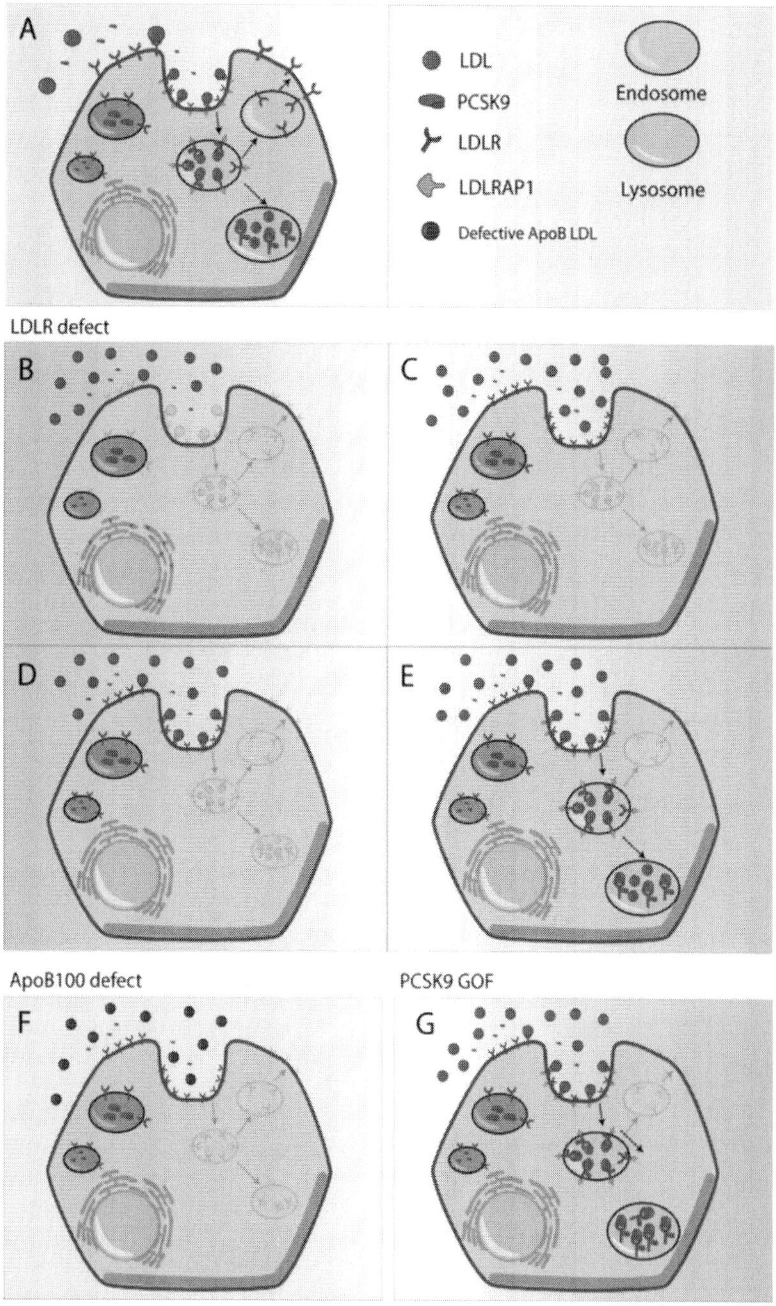

Homeostasis also entails Reverse Cholesterol Transport or RCT where cholesterol from cells with high cholesterol like macrophage foam cells is transported to the liver via HDLs for excretion via the feces. This entails the initial efflux of cholesterol from such cells with scavenger receptor-mediated endocytosis allowing oxLDL (Oxidized LDL) uptake by macrophages where cholesteryl ester-rich lipid droplets are the outcome by the action of acyl-coenzyme A:cholesterol acyltransferase. Free cholesterol is generated by lysosomal acid lipase or cytosolic cholesteryl ester hydrolase that is then subjected to efflux by:

- simple diffusion
- facilitated diffusion via scavenger receptor B1 (SR-BI)
- ATP binding cassette (ABC)A1 or ABCG1 transported mediated efflux

The ABCA1 pathway has received the limelight given that its deficiency results in Tangier disease. The oxysterols (LXR α and β ligands) of oxLDL upon cholesterol loading activate genes of the RCT pathway inclusive of ABCA1 that marks the start of the pathway. Apart from lipid-poor apoA-I, apoE has also been proposed to serve as the ABCA1-effluxed cholesterol from the plasma membrane. The free cholesterol in the apoprotein-effluxed lipids in nascent HDL is converted to cholesterol esters by the action of lecithin:cholesterol acyltransferase (LCAT). This forms the mature HDL from the nascent discoid. These mature HDLs can be taken up by the liver by the LDL receptor once the cholesterol in the HDL is transferred to apoB-containing lipoproteins by cholesteryl ester transfer protein (CETP). SR-BI or the LDL receptor family can also mediate cholesterol uptake. The cholesterol is transferred to bile as free cholesterol or bile acids to be excreted in feces. There is also the trans-intestinal cholesterol excretion (TICE), the non-biliary route where ABCG5/G8 uptakes cholesterol from plasma lipoproteins for excretion.

(Getz GS, Reardon CA, 2018; Afonso MS, Machado RM, Lavrador MS, QuintaoECR, Moore KJ, Lottenberg AM, 2018)

Figure 11 below depicts Reverse Cholesterol Transport or RCT: ***TOP***: Delivery of cholesterol and oxysterol by oxLDLto macrophages via scavenger receptors is involved in atherogenesis. ABCA1 and ABCG1 are induced by LXR ligands oxysterols with cholesterol efflux to apoA-I mediated by ABCA1and interactions with HDL are promoted by ABCG1. LCAT then causes esterification of lipid-poor apoA-I. Delivery to the cholesterol entails

CETP or SR-BI where it is then secreted in the bile. TICE entails enterocyte-mediated cholesterol uptake from lipoproteins to be excreted into the lumen by ABCG5/G8 (Afonso MS, Machado RM, Lavrador MS, QuintaoECR, Moore KJ, Lottenberg AM, 2018: © 2018 by the authors. Licensee MDPI, Basel, Switzerland. *This article is an open access article distributed under the terms and conditions of the Creative Commons Attribution (CC BY) license: http://creativecommons.org/licenses/by/4.0/*):

Key: FC: free cholesterol; CE: cholesterol ester; TG: triglycerides; CETP: cholesteryl ester transfer protein; LCAT: lecithin: cholesterol acyltransferase; TICE: trans-intestinal cholesterol excretion

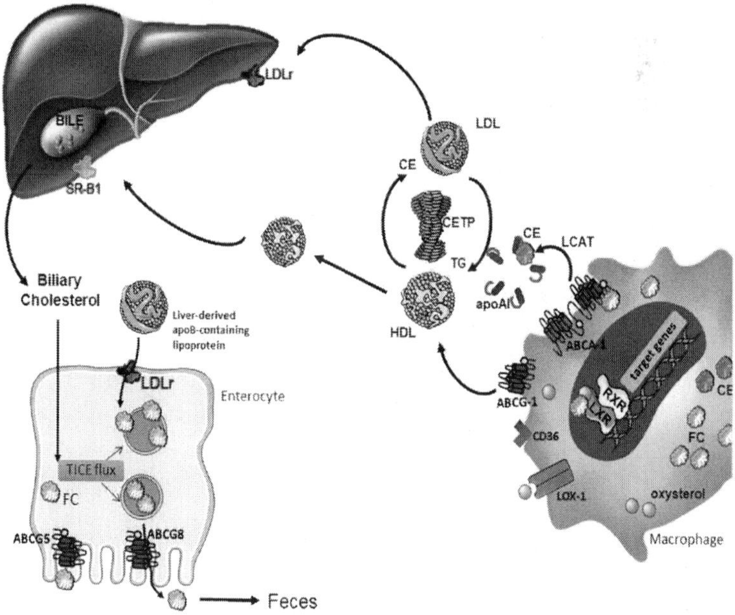

BOTTOM: Cholesterol uptake by apoprotein E (apoE) from cholesterol-loaded macrophages as unveiled by exogenous and endogenous production to form nascent discoidal high-density lipoprotein (HDL) that is later processed into mature spherical HDL particles in the plasma. Subsequent liver delivery entails scavenger receptor B1 (SR-BI) or low-density lipoprotein (LDL) receptor family or CETP(Getz GS, Reardon CA, 2018: © 2018 by the authors. Licensee MDPI, Basel, Switzerland. *This article is an open access article distributed under the terms and conditions of the Creative Commons Attribution (CC BY) license: http://creativecommons.org/licenses/by/4.0/):*

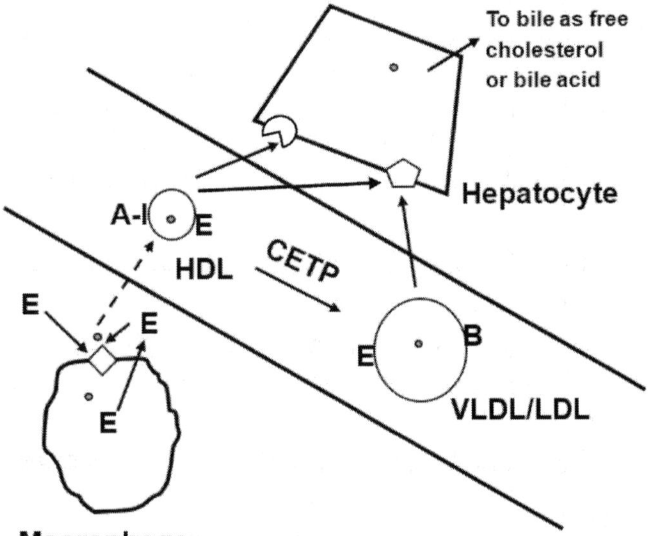

(Afonso MS, Machado RM, Lavrador MS, QuintaoECR, Moore KJ, Lottenberg AM, 2018;

Benito-Vicente A, Uribe KB, Jebari S, Galicia-Garcia U, Ostolaza H, Martin C, 2018;

Huff, Boyd and Jialal, 2021; Craig, Yarrarapu, Dimri, 2021)

1.3 CHOLESTEROL LEVELS

A high lipid profile or hyperlipidemia is associated with increased atherosclerotic cardiovascular disease (CVD) risks. The lipid panel is inclusive of:

- Total cholesterol
- High-density lipoprotein (HDL) cholesterol
- Low-density lipoprotein (LDL) cholesterol
- Triglycerides

The estimation of LDL-C = (total cholesterol) – (high-density lipoprotein cholesterol [HDL-C]) – (triglycerides/5) as per the Friedewald equation has been employed for long with the ratio of fasting triglyceride level(up to 4.5mmol/l) to very low-density lipoprotein cholesterol(TG: VLDL-C) kept at five. However, research is ongoing to overcome the challenges presented by this approach such as inaccurate estimation in hypertriglyceridemia and underestimation of VLDL remnants and IDLs.

According to the United States National Heart, Lung, and Blood Institute (NHLBI) expert panel (2011), screening children for dyslipidemia is warranted for familial hypercholesterolemia (FH). This entails the fasting lipid test for children with cardiovascular risks inclusive of diabetes, obesity, hypertension and family history. When there is no known cardiovascular risk, then the non-fasting lipid test is recommended. For adults, the conditions for screening are:

- Screening due to a family history of lipid disorders like familial hypercholesterolemia
- Probe the risk of 10-year cardiovascular diseases, such as the 2008 Framingham general cardiovascular risk score or JBS3 risk score
- Management of atherosclerotic cardiovascular diseases
- Evaluation for lipid-lowering therapy
- Pancreatitis

The major disorders of lipid metabolism or dyslipidemia can be primary like FH, familial dysbetalipoproteinemia, familial combined hyperlipidemia and chylomicronemia or secondary due to obstructive liver diseases, diabetes mellitus, chronic renal failure, hypothyroidism and drugs (those that lower HDL-C like androgens, anabolic steroids and beta-blockers or those that elevate LDL-C levels like phenothiazines, cyclosporine A and retinoids). These dyslipidemias can be high total cholesterol (TC), High low-density lipoprotein cholesterol (LDL-C), High non-high-density lipoprotein cholesterol (non-HDL-C), High triglycerides (TG), and Low high-density lipoprotein cholesterol (HDL-C).

The Expert Panel on Detection, Evaluation, and Treatment of High Blood Cholesterol in Adults. Executive Summary of The Third Report of The National Cholesterol Education Program (NCEP) Expert Panel on Detection, Evaluation, And Treatment of High Blood Cholesterol In Adults (Adult Treatment Panel III) lists the following recommendations for the same:

- "Fasting triglyceride level:
  - Normal: less than 150 mg/dL
  - Mild hypertriglyceridemia: 150 to 499 mg/dL
  - Moderate hypertriglyceridemia: 500 to 886 mg/dL
  - Very high or severe hypertriglyceridemia: greater than 886 mg/dL

- LDL-C level:
  - Optimal: less than 100 mg/ dL
  - Near optimal/above optimal:100 to 129 mg/dL
  - Borderline high: 130 to 159 mg/dL
  - High: 160 to 189 mg/dL
  - Very high: greater than 190 mg/dL
- HDL level:
  - Low: less than 40
  - High: greater than or equal to 60"

(Lee Y, SiddiquiWJ, StatPearls Publishing, 2021)

This becomes significant as the lipid profile is implicated in increasing the risk of atherosclerotic cardiovascular diseases(ASCVD) with the factors listed below:

"10-year risk cardiovascular assessment:

- Age (between 20 to 79)
- Sex
- Race
- Systolic blood pressure
- Diastolic blood pressure
- Total cholesterol
  1. LDL cholesterol
  2. HDL cholesterol
- History of diabetes
- On hypertension treatment?
- On a statin?
- On aspirin therapy?

The 10-year risk for ASCVD falls into the following categories:

- Low-risk (less than 5%)
- Borderline risk (5% to 7.4%)
- Intermediate risk (7.5% to 19.9%)
- High risk (greater than or equal to 20%)"
- (Lee Y, SiddiquiWJ, StatPearls Publishing, 2021)

With this background let us take up interesting and novel insights on cholesterol.

"The bile salts secreted by the gallbladder, which are needed for the digestion of fat and, therefore, the absorption of fat-soluble vitamins like A, D, and K, are made of cholesterol. Having a low cholesterol level in the body would therefore compromise a person's ability to digest fat."

— David Perlmutter, Grain Brain: The Surprising Truth about Wheat, Carbs, and Sugar--Your Brain's Silent Killers

1.4 REFERENCES

1. Daniel Steinberg, Thematic review series: The Pathogenesis of Atherosclerosis. An interpretive history of the cholesterol controversy: part I1, Journal of Lipid Research, Volume 45, Issue 9, 2004, Pages 1583-1593, ISSN 0022-2275, https://doi.org/10.1194/jlr.R400003-JLR200.
2. Steinberg D. The pathogenesis of atherosclerosis. An interpretive history of the cholesterol controversy, part IV: the 1984 coronary primary prevention trial ends it--almost. J Lipid Res. 2006 Jan;47(1):1-14. doi: 10.1194/jlr.R500014-JLR200. Epub 2005 Oct 15. PMID: 16227628.
3. Steinberg D. Thematic review series: the pathogenesis of atherosclerosis. An interpretive history of the cholesterol controversy, part V: the discovery of the statins and the end of the controversy. J Lipid Res. 2006 Jul;47(7):1339-51. doi: 10.1194/jlr.R600009-JLR200. Epub 2006 Apr 3. PMID: 16585781.
4. Craig M, YarrarapuSNS, Dimri M. Biochemistry, Cholesterol. [Updated 2021 Aug 18]. In: StatPearls [Internet]. Treasure Island (FL): StatPearls Publishing; 2022 Jan-. Available from: https://www.ncbi.nlm.nih.gov/books/NBK513326/
5. Huff T, Boyd B, Jialal I. Physiology, Cholesterol. [Updated 2021 Mar 2]. In: StatPearls [Internet]. Treasure Island (FL): StatPearls Publishing; 2022 Jan-. Available from: https://www.ncbi.nlm.nih.gov/books/NBK470561/
6. Benito-Vicente A, Uribe KB, Jebari S, Galicia-Garcia U, Ostolaza H, Martin C. Familial Hypercholesterolemia: The Most Frequent Cholesterol Metabolism Disorder Caused Disease. International Journal of Molecular Sciences. 2018; 19(11):3426. https://doi.org/10.3390/ijms19113426
7. Afonso MS, Machado RM, Lavrador MS, QuintaoECR, Moore KJ, Lottenberg AM. Molecular Pathways Underlying Cholesterol Homeostasis. Nutrients. 2018; 10(6):760. https://doi.org/10.3390/nu10060760
8. Charles KN, Shackelford JE, Faust PL, FlieslerSJ, Stangl H and Kovacs WJ (2020) Functional Peroxisomes Are Essential for Efficient Cholesterol Sensing and Synthesis. Front. Cell Dev. Biol. 8:560266. doi: 10.3389/fcell.2020.560266

9. Zhou F and Sun X (2021) Cholesterol Metabolism: A Double-Edged Sword in Hepatocellular Carcinoma. Front. Cell Dev. Biol. 9:762828. doi: 10.3389/fcell.2021.762828

10. Dong, XY., Tang, SQ. & Chen, JD. Dual functions of Insig proteins in cholesterol homeostasis. Lipids Health Dis 11, 173 (2012). https://doi.org/10.1186/1476-511X-11-173

11. Getz GS, Reardon CA. Apoprotein E and Reverse Cholesterol Transport. International Journal of Molecular Sciences. 2018; 19(11):3479. https://doi.org/10.3390/ijms19113479

12. Lee Y, SiddiquiWJ. Cholesterol Levels. [Updated 2021 Jul 26]. In: StatPearls [Internet]. Treasure Island (FL): StatPearls Publishing; 2022 Jan-. Available from: https://www.ncbi.nlm.nih.gov/books/NBK542294/

CHAPTER

2 Trends In Cholesterol: Research Insights From Global And Local Studies

"Cholesterol is a substance in the blood that causes you to eat salads"

– Author: Richard Carleton

Contents

2.1 NON-COMMUNICABLE DISEASE RISK FACTOR COLLABORATION (NCD-RISC) HIGHLIGHTS CHANGES

The cholesterol profile of 82.1 million participants of 23 Asian and Western countries of 458 population-based studies was probed by the Non-Communicable Disease Risk Factor Collaboration (NCD-RisC): a collaboration to gauge the risk factors of NCD globally. The article published in the *International Journal of epidemiology* entailed the analyses of total cholesterol (TC), HDL and non-HDL cholesterol and the total-to-HDL cholesterol ratio for ages 40–79 years and the timespan 1970–2018 for:

- Nordic countries: Finland, Iceland and Norway.
- Eastern Central Europe: Czech Republic, Lithuania, Poland and Slovakia.
- Western central Europe: Belgium, Germany and Switzerland.
- Southern Europe: France, Italy and Spain.
- High-income English-speaking countries: Australia, Canada, New Zealand, UK and USA.
- East and Southeast Asia: China, Japan and South Korea.

Sweden and Thailand were included for TC analyses only on the basis of data availability.

Looking at TC, 40–59 year-olds of both genders of most western nations documented a dip (except Poland) with Nordic countries and central Europe documenting a conspicuous lowering. All four Asian countries documented an increase in TC with China and Thailand emerging the most increases. An increase in mean HDL cholesterol accounted for the elevations in TC in Japan and South Korea while this increase in China was attributed to non-HDL cholesterol. The lowering of TC in Western countries was attributed to increased HDL and lowered non-HDL (except in Germany and Norway as well as and men in the Czech Republic and Slovakia).

Figure 1 below illustrates the changes in cholesterol across 23 Asian and Western countries with 82.1 million participants: ***TOP***: mean total cholesterol (NCD Risk Factor Collaboration (NCD-RisC) © The Author(s) 2019. Published by Oxford University Press on behalf of the International Epidemiological Association. *This is an Open Access article distributed under the terms of the Creative Commons Attribution License (http://creativecommons.org/licenses/by/4.0/), which permits unrestricted reuse,* *distribution, and reproduction in any medium, provided the original work is properly cited*):

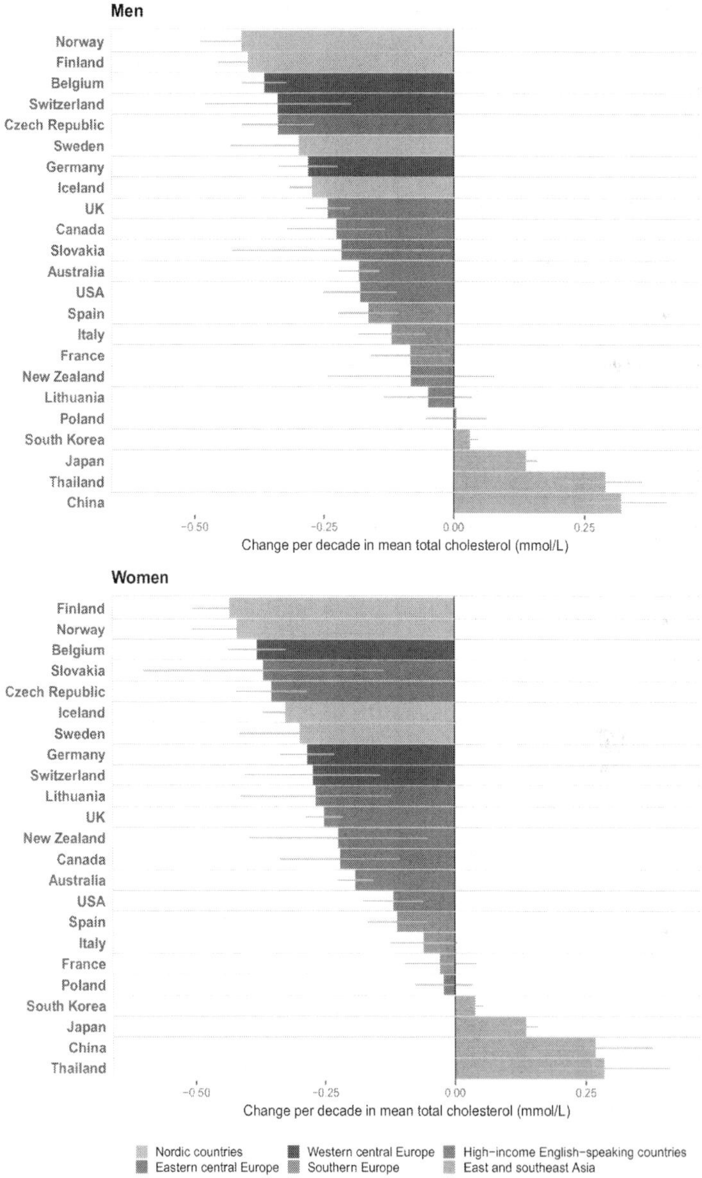

BOTTOM: The changes in (A) HDL and (B) non-HDL cholesterol (NCD Risk Factor Collaboration (NCD-RisC) © The Author(s) 2019. Published by Oxford University Press on behalf of the International Epidemiological Association. *This is an Open Access article distributed under the terms*

of the Creative Commons Attribution License (http://creativecommons.org/licenses/by/4.0/), which permits unrestricted reuse, distribution, and reproduction in any medium, provided the original work is properly cited):

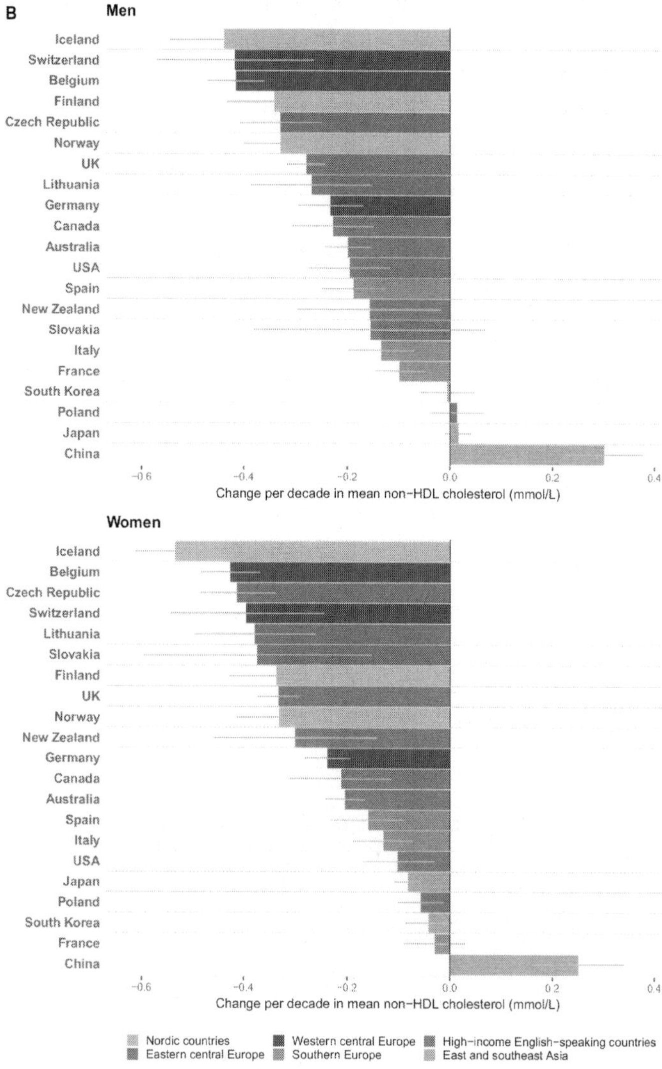

Continued.

The mean total-to-HDL cholesterol ratio documented an increase in China while most Western countries showed a decrease in this aspect. Figure 2 below illustrates ***TOP***: mean total-to-HDL cholesterol ratio change(NCD Risk Factor Collaboration (NCD-RisC) © The Author(s) 2019. Published by Oxford University Press on behalf of the International Epidemiological

Association. *This is an Open Access article distributed under the terms of the Creative Commons Attribution License (http://creativecommons.org/licenses/by/4.0/), which permits unrestricted reuse, distribution, and reproduction in any medium, provided the original work is properly cited*):

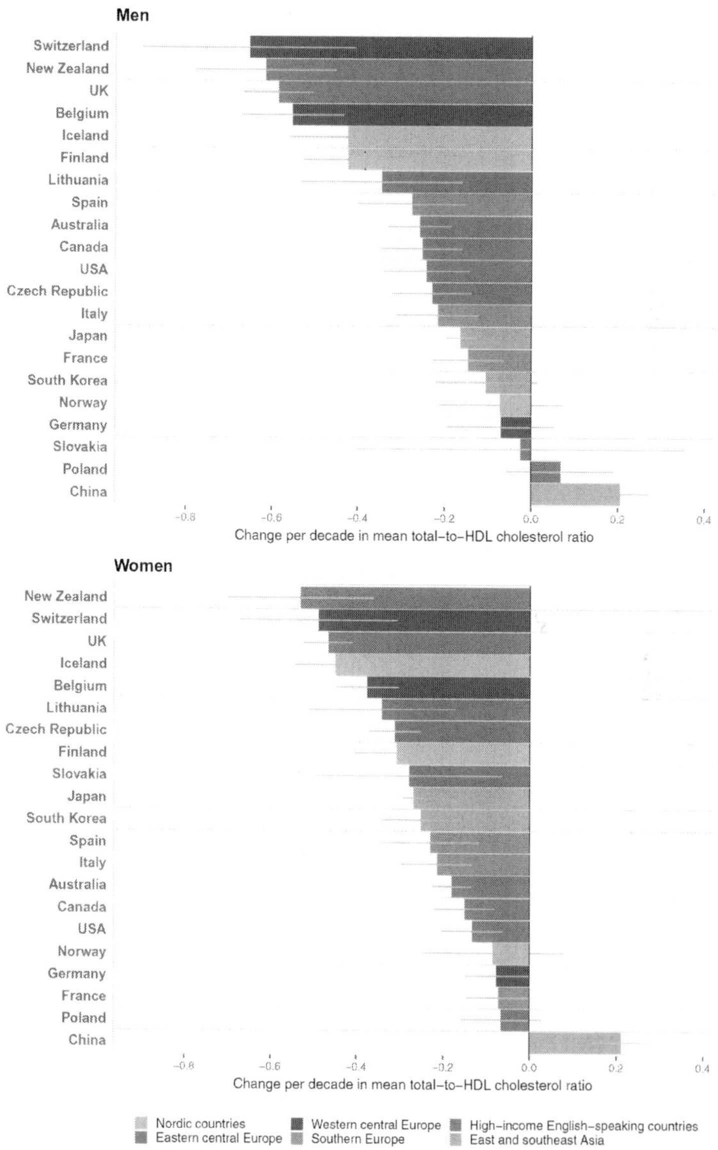

BOTTOM: Change in mean (A) total cholesterol, (B) non-HDL cholesterol, (C) HDL cholesterol and (D) total-to-HDL cholesterol ratio for age groups 40–59 vs. 60–79 years where the changes documented moderate to strong

Further analyses unveiled that truncal obesity over generalized obesity documented an evident correlation with non-HDL cholesterol, remnant cholesterol, triglycerides and the total:HDL cholesterol ratio (two-line regression analysis; $p < 0.05$). Table 1 below documents this regression analysis of the aberrant cholesterol profile with obesity (BMI ≥ 25 kg/m2) and truncal obesity (waist: hip > 0.9) (Gupta, R, Guptha, S, Agrawal, A et al, 2008: *This article is published under license to BioMed Central Ltd. This is an Open Access article distributed under the terms of the Creative Commons Attribution License (http://creativecommons.org/licenses/by/2.0), which permits unrestricted use, distribution, and reproduction in any medium, provided the original work is properly cited*):

Lipid variable	Educational status	Obesity	Truncal obesity
Total Cholesterol ≥ 200 mg/dl	0.450 (0.072)	0.322 (0.120)	0.773 (0.011)
LDL Cholesterol ≥ 130 mg/dl	0.529 (0.051)	0.462 (0.069)	0.737 (0.014)
HDL Cholesterol < 40 mg/dl	0.347 (0.109)	0.352 (0.107)	0.477 (0.064)
Non-HDL Cholesterol ≥ 160 mg/dl	0.531 (0.050)	0.421 (0.082)	0.884 (0.003)
Remnant cholesterol ≥ 25 mg/dl	-0.211 (0.179)	-0.257 (0.152)	-0.0711 (0.305)
Total:HDL cholesterol ratio ≥ 5.0	0.469 (0.067)	0.456 (0.071)	0.593 (0.037)
Total:HDL cholesterol ratio ≥ 4.0	0.535 (0.049)	0.480 (0.061)	0.768 (0.011)
Triglycerides ≥ 150 mg/dl	-0.111 (0.259)	-0.196 (0.189)	0.099 (0.272)

A 2014-published article by Joshi and team (2014) funded by the Indian Council of Medical Research, New Delhi (No. 55/1/TF/Diab./07-NCD-II) scrutinized the lipid profile of Phase I of the Indian Council of Medical Research India Diabetes Study (ICMR-INDIAB study). This encompassed 20–90-year-old individuals from three states that were randomly representative of the south (Tamilnadu), west (Maharashtra), and east (Jharkhand) of India along with one union territory (UT) representing northern India (Chandigarh).

The definitions entailed in the study as presented by the authors are listed below (Joshi SR, Anjana RM, Deepa M, Pradeepa R, Bhansali A, Dhandania VK, et al, 2014: © 2014 Joshi et al. *This is an open-access article distributed under the terms of the Creative Commons Attribution License, which permits unrestricted use, distribution, and reproduction in any medium, provided the original author and source are credited*):

"***Dyslipidemia**: National Cholesterol Education Programme (NCEP) guidelines [10]* were used for definition of dyslipidemia *as follows:*

Hypercholesterolemia – serum cholesterol levels ≥200 mg/dl (≥5.2 *mmol/l).*

Hypertriglyceridemia – serum triglyceride levels ≥150 mg/dl (≥1.7 *mmol/l).*

Low HDL cholesterol – HDL cholesterol levels <40 mg/dl (<1.04 *mmol/l) for men and <50 mg/dl (<1.3 mmol/l) for women.*

High LDL cholesterol – LDL cholesterol levels ≥130 mg/dl (≥3.4 *mmol/l) calculated using the Friedewald equation.*

High total cholesterol to HDL-C ratio*: This is defined as a total cholesterol to HDL-C ratio of ≥4.5.*

Isolated hypercholesterolemia*: Serum cholesterol ≥200 mg/dl and triglycerides <150 mg/dl;* ***Isolated hypertriglyceridemia****: Serum triglycerides ≥150 mg/dl and cholesterol <200 mg/dl;* ***Isolated low HDL-C****: HDL-C ≤40 mg/dl (male) and ≤50 mg/dl (female) without hypertriglyceridemia or hypercholesterolemia.*

Diabetes*: Individuals diagnosed by a physician and on antidiabetic medications (self-reported) and/or those who had fasting CBG ≥126 mg/dl (≥7 mmol/L) and/or 2-hr post-glucose CBG value ≥220 mg/dl (≥12.2 mmol/L) [11].*

Impaired fasting glucose [IFG]*: Fasting CBG ≥110 mg/dl (≥6.1 mmol/L) and <126 mg/dl (<7 mmol/L) and 2-hr post-glucose value <160 mg/dl (<8.9 mmol/L) [11].*

Impaired glucose tolerance [IGT]*: Two-hour post-glucose CBG ≥160 mg/dl (≥8.9 mmol/L) but <220 mg/dl (<12.2 mmol/L) and fasting value <126 mg/dl (<7 mmol/L) [11].*

Prediabetes*: Individuals with IFG or IGT or both.*

Dysglycemia*: Presence of diabetes and / or prediabetes.*

Hypertension*: Individuals diagnosed by a physician and on antihypertensive medications (self-reported) and/or those who had systolic blood pressure ≥140 mmHg and/or diastolic blood pressure ≥90 mmHg – Joint National Committee 7 (JNC7) Criteria [12].*

Obesity*: Generalized obesity was defined as BMI ≥25 kg/m^2; overweight as BMI 23–25 kg/m^2* and abdominal obesity was defined as waist ≥90 cm (males), ≥80 cm (females) using Asia-Pacific guidelines for south Asians *[13].*

Coronary Artery Disease (CAD)*: CAD was diagnosed based on positive medical history (documented myocardial infarction (MI), angina pectoris* *and coronary artery bypass graft) and/or ischemic changes on a conventional 12-lead ECG which included ST-segment depression (Minnesota codes 1-1-1 to 1-1-7) or Q-wave changes (Minnesota codes 4–1 to 4–2) [14]"*

The authors unveiled that 13.9% had hypercholesterolemia, 29.5% documented hypertriglyceridemia, 72.3% showed decreased HDL-C, 11.8% displayed elevated LDL-C levels while abnormalities in one of the lipid parameters were seen in 79% of the participants. The lowered HDL-C emerged as the most predominant lipid abnormality with all abnormalities detected more in urban dwellers. The risk factors for the aberrations emerged as obesity, diabetes, and dysglycemia.

Table 2 below documents the findings of the researchers across the four zones -(Joshi SR, Anjana RM, Deepa M, Pradeepa R, Bhansali A, Dhandania VK, et al, 2014: © 2014 Joshi et al. *This is an open-access article distributed under the terms of the Creative Commons Attribution License, which permits unrestricted use, distribution, and reproduction in any medium, provided the original author and source are credited*):

Prevalence	All	four	regions		Tamilnadu			Maharashtra			Jhar khand			Chandi garh	
	Urban	Rural	Overall	Urban	Rural	Overall	Urban	Rural	Overall	Urban	Rural	Overall	Urban	Rural	Overall
n	590	1452	2042	194	463	657	115	358	473	133	277	410	148	354	502
Hypercholesterolemia	19.2	11.7**	13.9	23.7	16.0*	18.3	13.9	13.7	13.7	10.5	2.2**	4.9	25.0	11.6**	15.5
Hypertriglyceridemia	36.6	26.6**	29.5	33.0	29.6	30.6	28.7	20.9	22.8	38.3	17.3**	24.1	45.9	35.6*	38.6
Low HDL cholesterol	73.2	71.9	72.3	72.2	67.6	68.9	69.6	69.6	69.6	74.4	78.0	76.8	76.4	75.1	75.5
High LDL cholesterol	15.9	10.1**	11.8	22.2	13.2*	15.8	12.2	13.7	13.3	7.5	1.4*	3.4	18.2	9.3*	12.0
High total cholesterol: HDL ratio	44.2	33.5**	36.6	48.5	39.7*	42.3	33.9	34.6	34.5	38.3	19.1**	25.4	52.0	35.6**	40.4
Isolated hypercholesterolemia	4.6	4.5	4.6	5.7	5.8	5.8 '	5.2	6.7	6.3	3.0	1.1	1.7	4.1	3.4	3.6
Isolated hypertriglyceridemia	22.0	19.4	20.2	14.9	19.4	18.1	20.0	14.0	15.4	30.8	16.2*	21.0	25.0	27.4	26.7
Isolated low HDL-C	39.0	47.2*	44.9	39.2	41.0	40.5	41.7	49.2	47.4	39.1	61.7**	54.4	36.5	42.1	40.4

*p<0.05 and **p<0.001 compared to urban participants; Dyslipidemia was diagnosed using NCEP guidelines – hypercholesterolemia: Total cholesterol (TC) ≥200 mg/dl; hypertriglyceridemia: triglycerides ≥150 mg/dl; low HDL cholesterol: HDL-C <40 (males) and <50 mg/dl (females); high LDL cholesterol: LDL-C ≥130 mg/dl; High total cholesterol: HDL-C ratio ≥4.5; Isolated hypercholesterolemia: TC ≥200 mg/dl and triglycerides <150 mg/dl; Isolated hypertriglyceridemia: serum triglycerides ≥150 mg/dl and TC <200 mg/dl; Isolated low HDL-C: HDL-C ≤40 mg/dl (male) and ≤50 mg/dl (female) without hypertriglyceridemia or hypercholesterolemia.
doi:10.1371/journal.pone.0096808.t002

Variables	Tamilnadu		Maharashtra		Jharkhand		Chandigarh	
	No lipid abnormality (n = 152)	Subjects with any lipid abnormality (n = 505)	No lipid abnormality (n = 109)	Subjects with any lipid abnormality (n = 364)	No lipid abnormality (n = 81)	Subjects with any lipid abnormality (n = 329)	No lipid abnormality (n = 86)	Subjects with any lipid abnormality (n = 416)
Age (years)	41.2±14.9	42.7± 14.1	40.9±15.3	41.9±13.8	40.0±14.3	37.5±13.0	33.7±11.2	36.2±12.1
Male n(%)	103 (67.8)	229 (45.3)**	63 (57.8)	179 (49.2)	58 (71.6)	163 (49.5)**	53 (61.6)	223 (53.6)
Body mass index (kg/m²)	20.5±3.6	22.7±4.3**	19.7±3.6	21.6±4.0**	19.0±2.3	21.0±4.2**	21.5±3.7	24.0±4.8**
Waist circumference (cm)								
Male	75.6±10.2	82.8±12.2**	75.2±10.9	79.8±10.4**	73.4±6.9	80.2±13.1*	77.6±10.7	85.7±12.5**
Female	69.9±11.1	75.8±11.4*	63.9±10.4	71.8±11.3**	68.6±9.7	72±11.7	73.5±10.4	80.3±13.4
Systolic blood pressure (mm Hg)	134±24	138±22	122±14	133±20*	132±16	132±17	125±10	128±18
Diastolic blood pressure (mm Hg)	78±15	83±13	73±7	81±10*	79±12	81±9	80±10	78±13
Serum cholesterol (mg/dl)	157±22	170±44*	155±22	162±39	136±25	138±38	151±23	162±41*
Serum triglycerides (mg/dl)	82	130**	82	118**	82	113**	89	142**
HDL cholesterol (mg/dl)	52±9	36±9**	54±13	37±11**	50±8	34±8**	51±10	36±9**
LDL cholesterol (mg/dl)	89±21	103±34**	84±23	96±34*	69±22	77±29*	81±22	92±33*
Total cholesterol: HDL ratio	3.1±0.6	5±2.7**	3.0±0.7	4.7±2.1**	2.8±0.5	4.2±1.4**	30.0±0.6	4.7±1.5**
Current smoking n(%)	13 (19.7)	83 (16.4)	14 (12.8)	27 (7.4)	11 (13.6)	28 (8.5)	17 (19.8)	71 (17.1)
Current alcohol users n(%)	40 (26.3)	82 (16.2)*	9 (8.3)	32 (8.8)	33 (40.7)	76 (23.1)*	16 (18.8)	54 (13.0)
Education level n(%)								
Illiterate	43 (28.3)	144 (28.5)	27 (24.8)	107 (29.4)	27 (33.3)	148 (45.0)	18 (20.9)	69 (16.6)
Primary, middle or high schooling	101 (66.4)	325 (64.4)	74 (67.9)	229 (62.9)	44 (54.3)	158 (48.0)	57 (66.3)	297 (71.4)
Graduate and above	8 (5.3)	36 (7.1)	8 (7.3)	28 (7.7)	10 (12.3)	23 (7.0)	11 (12.8)	50 (12.0)
Occupation n(%)								
Professional	4 (3.3)	22 (4.8)	3 (3.1)	13 (3.8)	10 (15.6)	15 (5.1)	5 (6.8)	19 (5.1)
Labourer	74 (61.7)	255 (55.2)	63 (64.9)	200 (58.7)	29 (45.3)	124 (42.2)	40 (54.1)	157 (41.9)
Others	42 (35.0)	185 (40.0)	31 (32.0)	128 (37.5)	25 (39.1)	155 (52.7)	29 (39.2)	199 (53.1)
Income level n(%)								
<5,000 INR	124 (83.2)	393 (79.4)	82 (75.2)	259 (72.8)	59 (77.6)	208 (77.0)	26 (31.7)	166 (43.5)
5,000–10,000 INR	14 (9.4)	76 (15.4)	15 (13.8)	65 (18.3)	13 (17.1)	27 (10.0)	33 (40.2)	105 (27.5)
>10,000 INR	11 (7.4)	26 (5.3)	12 (11.0)	32 (9.0)	4 (5.3)	35 (13.0)	23 (28.0)	111 (29.1)
Self reported diabetes n(%)	5 (3.3)	33 (6.5)	1 (0.9)	14 (3.8)	1 (1.2)	8 (2.4)	0	24 (5.8)
Coronary artery disease n(%)	5 (3.3)	29 (5.7)	3 (2.8)	17 (4.7)	2 (2.5)	9 (2.7)	0	10 (2.4)
Family history of diabetes n(%)	17 (11.2)	88 (17.4)	2 (1.8)	24 (6.6)	3 (3.7)	22 (6.7)	7 (8.1)	43 (10.3)

*p<0.05 and **p<0.001 compared to subjects with no lipid abnormality.
doi:10.1371/journal.pone.0096808.t001

2.3 STUDIES FROM THE US

The profile of LDL-C of 105 million patients from 2001 to 2011 described as "*the largest nationally-based retrospective study of LDL-C trends reported to-date*" by Kaufman and team of Quest Diagnostics and Celera (a Quest Diagnostics company) revealed that the average rate of annual decline was −2.05 mg/dL (95% CI [−2.35, −1.75]) between 2001 and 2008 after which a flattening out of values was recorded: 104.5 mg/dL in 2009 and 104.3 mg/dL in 2011 for LDL-C. These results were obtained for all age groups and genders entailing 195,050,298 observations in the eleven-year timespan.

Figure 4 below depicts ***TOP***: The trend in LDL-C across the study between 2001 and 2011 (© 2013 Kaufman et al. *This is an open-access article distributed under the terms of the Creative Commons Attribution License, which permits unrestricted use, distribution, and reproduction in any medium, provided the original author and source are credited*):

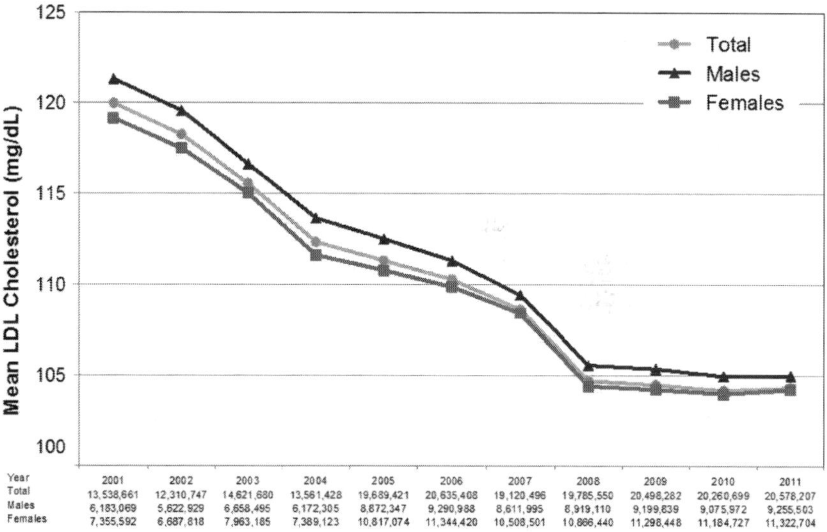

BOTTOM: LDL-C trends by age for the study between 2001 and 2011(© 2013 Kaufman et al. *This is an open-access article distributed under the terms of the Creative Commons Attribution License, which permits unrestricted use, distribution, and reproduction in any medium, provided the original author and source are credited*):

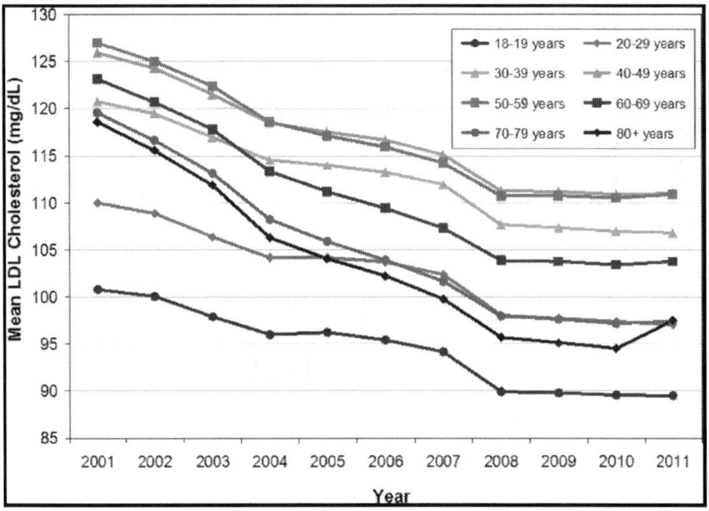

Researchers Aranmolate and Obayemi (2018) probed the Behavioral Risk Factor Surveillance System (BRFSS)-2015 data encompassing 18 year-old-non-institutionalized US individuals (n=

5039; 1703 male and 3336 females). It emerged that elderly and middle-age adults documented increased cholesterol screening as opposed to young adults in Mississippi. This lowered screening was also documented by those with less education. These findings become pertinent calling for efficient cholesterol awareness programs.

Table 3 below documents the high cholesterol trends in Mississippi Behavioral Risk Factor Surveillance System(BRFSS-2015) participants as presented by the researchers (Aranmolate and Obayemi, 2018: Copyright © 2018 by Author/s and Licensed by Lectito BV, Netherlands. *This is an open access article distributed under the Creative Commons Attribution License which permits unrestricted use, distribution, and reproduction in any medium, provided the original work is properly cited*):

TOP: Those who were aware they had high cholesterol (Aranmolate and Obayemi, 2018: Copyright © 2018 by Author/s and Licensed by Lectito BV, Netherlands. *This is an open access article distributed under the Creative Commons Attribution License which permits unrestricted use, distribution, and reproduction in any medium, provided the original work is properly cited*):

VARIABLES	YES			NO		
	Percent (%)	95% CI	N	Percent (%)	95% CI	N
Gender						
Male	38.6	35.6-41.6	813	64.4	58.4-64.4	947
Female	38.9	36.6-41.2	1600	61.1	58.8-63.5	1839
Age Group						
18-24	N/A	N/A	N/A	92.9	87.8-98.0	88
25-34	12.9	8.5-17.2	41	87.1	82.8-91.5	266
35-44	27.7	22.9-32.5	135	72.3	67.5-77.1	358
45-54	38.4	34.1-42.8	298	61.6	57.2-66.0	476
55-64	54.6	50.9-58.3	659	45.4	41.7-49.1	617
65+	57.6	54.8-60.3	1271	42.4	39.7-45.2	981
Race						
White, non-Hispanic	41.9	39.6-44.2	1605	58.1	55.8-60.5	1720
Black, non-Hispanic	34.2	31.0-37.4	769	65.8	62.9-69.0	1009
AI/AN, non-Hispanic	N/A	N/A	N/A	N/A	N/A	N/A
Asian, non-Hispanic	N/A	N/A	N/A	N/A	N/A	N/A
Native Hawaiian, non-Hispanic	N/A	N/A	N/A	N/A	N/A	N/A
Other, non-Hispanic	N/A	N/A	N/A	N/A	N/A	N/A
Multiracial, non-Hispanic	N/A	N/A	N/A	N/A	N/A	N/A
Hispanic	N/A	N/A	N/A	N/A	N/A	N/A
Level of Education						
Less than H.S	52.2	46.7-57.8	374	47.8	42.3-53.4	289
H.S or G.E.D	42.0	38.7-45.4	811	58.0	54.6-61.4	782
Some post-H.S	33.4	30.1-36.6	617	66.6	63.4-69.9	779
College Graduate	32.5	29.3-35.6	603	67.6	64.4-70.7	926
Total	38.7	36.9-40.6	2413	61.3	59.4-63.1	2786

Data Source: Behavioral Risk Factor Surveillance System, 2015

BOTTOM: Those who underwent cholesterol screening in the previous five years (Aranmolate and Obayemi, 2018: Copyright © 2018 by Author/s and Licensed by Lectito BV, Netherlands. *This is an open access article distributed under the Creative Commons Attribution License which permits unrestricted use, distribution, and reproduction in any medium, provided the original work is properly cited*):

VARIABLES	Checked in the past 5 years			Not Checked in Past 5 years			Never Checked		
	Percent (%)	95% CI	N	Percent (%)	95% CI	N	Percent (%)	95% CI	N
Gender									
Male	72.8	69.8-75.7	1703	2.5	1.6-3.4	46	24.8	21.8-27.7	295
Female	79.0	76.8-81.3	3336	1.9	1.3-2.6	65	19.0	16.8-21.2	383
Age Group									
18-24	40.2	32.8-47.6	89	N/A	N/A	N/A	57.3	50.1-65.0	123
25-34	59.2	53.9-64.6	287	N/A	N/A	N/A	38.3	33.0-43.6	167
35-44	77.2	72.9-81.5	467	3.2	1.4-5.1	16	19.6	15.5-23.6	106
45-54	82.9	79.6-86.2	739	2.7	1.4-4.0	27	14.4	11.3-17.5	102
55-64	91.0	88.8-93.1	1243	1.3	0.6-1.9	22	7.8	5.8-9.8	91
65+	94.2	92.7-95.8	2205	1.3	0.7-2.0	26	4.4	3.1-5.8	89
Race									
White, non-Hispanic	79.1	79.6-81.4	3203	3	2.1-3.8	91	17.9	15.7-20.1	354
Black, non-Hispanic	72.7	69.4-75.9	1744	N/A	N/A	N/A	26.3	23.1-29.5	293
AI/AN, non-Hispanic	N/A	N/A	N/A	N/A	N/A	N/A	N/A	N/A	N/A
Asian, non-Hispanic	N/A	N/A	N/A	N/A	N/A	N/A	N/A	N/A	N/A
Native Hawaiian, non-Hispanic	N/A	N/A	N/A	N/A	N/A	N/A	N/A	N/A	N/A
Other, non-Hispanic	N/A	N/A	N/A	N/A	N/A	N/A	N/A	N/A	N/A
Multiracial, non-Hispanic	N/A	N/A	N/A	N/A	N/A	N/A	N/A	N/A	N/A
Hispanic	N/A	N/A	N/A	N/A	N/A	N/A	N/A	N/A	N/A
Level of Education									
Less than H.S	69.1	63.7-74.5	642	N/A	N/A	N/A	29.2	23.8-34.6	113
H.S or G.E.D	73.2	70.0-76.5	1540	2.0	1.1-2.9	36	24.8	21.5-28.0	257
Some post-H.S	76.1	72.8-79.4	1352	2.7	1.6-3.8	34	21.2	18.0-24.4	186
College Graduate	86.9	84.4-89.4	1489	2.1	1.2-3.0	33	11.0	8.7-13.4	119
Total	76.0	74.2-77.9	5039	2.2	1.6-2.7	111	21.8	20.0-23.6	678

Data Source: Behavioral Risk Factor Surveillance System, 2015

2.4 THE SCENARIO FROM CHINA

Researchers Wang and team (2011) scrutinized the cholesterol profile in 251 Chinese individuals (45 to 89 years of age) from the population-based Beijing Eye Study, 2006. They reported that the prevalence of dyslipidemia was 56.1±0.9%% with total, HDL cholesterol, LDL cholesterol and triglycerides at 4.92±1.01 mmol/L, 1.61±0.36 mmol/L, 2.88±0.85 mmol/L, and 1.76±1.29 mmol/L, respectively. Of these, 50.9% of them were aware of this condition, 23.8% were treated, and 39.91% had the levels under control. Associations emerged for age (odds ratio (OR):1.02; 95% confidence interval (CI): 1.01, 1.03), female gender (OR:1.51; 95%CI: 1.25, 1.83), urban region (OR:1.82; 95%CI: 1.30, 2.55), body mass index (OR:1.13; 95%CI: 1.10, 1.15), income (OR:1.11; 95%CI:1.02, 1.21), blood glucose concentration (OR:1.10; 95%CI:1.05, 1.16), diastolic blood pressure (OR:1.02; 95%CI: 1.01, 1.03), and smoking (OR:1.23; 1.01, 1.51) with dyslipidemia that warrants a deeper probing of such links.

Figure 5 below illustrates ***TOP***: the cholesterol profile in the Beijing Eye Study, 2006 population assessed (Wang S, Xu L, Jonas JB, You QS, Wang YX, Yang H, 2011: © 2011 Wang et al. *This is an open-access article distributed under the terms of the Creative Commons Attribution License, which permits unrestricted use, distribution, and reproduction in any medium, provided the original author and source are credited*):

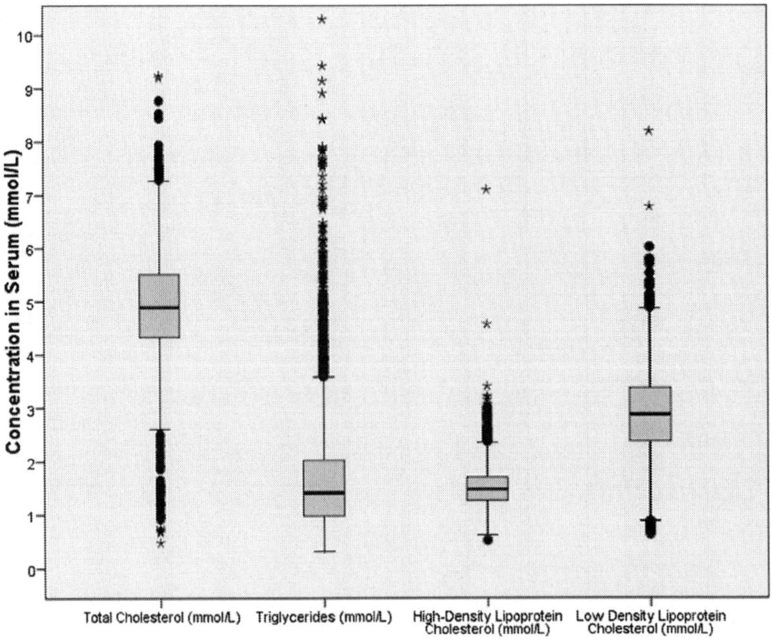

BOTTOM: The factors linked with aberrant cholesterol profiles in the Beijing Eye Study, 2006 population (Wang S, Xu L, Jonas JB, You QS, Wang YX, Yang H, 2011: © 2011 Wang et al. *This is an open-access article distributed under the terms of the Creative Commons Attribution License, which permits unrestricted use, distribution, and reproduction in any medium, provided the original author and source are credited*):

Univariate analysis:

Parameter	P-Value	Exponential B	95% Confidence Interval
Age	<0.001	1.02	1.02, 1.03
Female gender	0.001	1.29	1.11, 1.49
Urban region	<0.001	2.09	1.81, 2.43
Body mass index	<0.001	1.11	1.08, 1.13
Body height	0.02	0.99	0.98, 0.99
Body weight	<0.001	1.02	1.02, 1.03
Level of education	<0.001	1.15	1.08, 1.23
Income	<0.001	1.20	1.15, 1.24
Fasting serum concentration glucose concentration	<0.001	1.14	1.08, 1.20
Systolic blood pressure	0.02	1.01	1.01, 1.02
Mean arterial blood pressure	0.03	1.01	1.01, 1.02
Consumption of wine	0.001	0.86	0.78, 0.94
Smoking	0.002	0.79	0.67, 0.92
Consumption of beer	0.79		

Binary regression analysis:

Parameter	P-Value	Exponential B	95% Confidence Interval
Age	<0.001	1.02	1.01, 1.03
Gender	<0.001	1.51	1.25, 1.83
Urban region	0.001	1.82	1.30, 2.55
Body mass index	<0.001	1.13	1.10, 1.15
Income	0.01	1.11	1.02, 1.21
Blood concentr. of glucose	<0.001	1.10	1.05, 1.16
Diastolic blood pressure	0.02	1.02	1.01, 1.03
Smoking	0.04	1.23	1.01, 1.51
Level of education	0.40		
Systolic blood pressure	0.56		
Consumption of wine	0.24		

doi:10.1371/journal.pone.0017326.t001

Researchers Opoku, Gan and Fu (2019) reported a comparable prevalence of dyslipidemia in rural (43.2%) and urban (43.3%) areas in China in a study described as "*the first study to nationally compare rural and urban prevalence of dyslipidemia and risk factors among middle-aged and older Chinese adults (≥40 years)*". The study encompassed 136,945 individuals of the China National Stroke Screening and prevention project (CNSSPP) with the lipid profile documented in table 4 below (Opoku, S., Gan, Y., Fu, W. et al, 2019: *Open Access: This article is distributed under the terms of the Creative Commons Attribution 4.0 International License (http:// creativecommons.org/licenses/by/4.0/), which permits unrestricted use, distribution, and reproduction in any medium, provided you give appropriate credit to the original author(s) and the source, provide a link to the Creative*

Commons license, and indicate if changes were made. The Creative Commons Public Domain Dedication waiver (http://creativecommons.org/publicdomain/zero/1.0/) applies to the data made available in this article, unless otherwise stated):

Key: TC total cholesterol; *HDL-C* high-density lipoprotein; *LDL-C* low-density lipoprotein: triglycerides; *M/W; Men/ Women; BMI*[a] weight *in kilograms divided by height in meters squared:* [b,c] Two different geographic zones

	Rural (*N*=72,712)	**Urban (*N*=64,233)**	**Total (*N*=13,6945)**
	No. (%)	**No. (%)**	**No. (%)**
Age			
Median (IQR)	55.0 (49.0–62.0)	56.0 (50.0–63.0)	55.0 (49.0–63.0)
40–49	20,129 (27.7)	15,922 (24.8)	36,051 (26.3)
50–59	30,366 (41.8)	27,494 (42.8)	57,860 (42.3)
60–69	13,923 (19.1)	12,560 (19.6)	26,483 (19.3)
70 and above	8294 (11.4)	8257 (12.8)	16,552 (12.1)
Sex			
Men	37,728 (51.9)	32,835 (51.1)	70,563 (51.5)
Women	34,984 (48.1)	31,398 (48.9)	66,382 (48.5)
History of CHD			
Yes	65,704 (90.4)	58,702 (91.4)	12,4406 (90.8)
No	7008 (9.6)	5531 (8.6)	12,539 (9.2)
History of Dyslipidemia			
Yes	44,834 (61.8)	40,346 (64.2)	85,180 (62.9)
No	27,722 (38.2)	22,501 (35.8)	50,223 (37.1)
History of Stroke			
Yes	56,563 (77.8)	51,846 (80.7)	10,8409 (79.2)
No	16,150 (22.2)	12,387 (19.3)	28,537 (20.8)
Geographical Regions			
[b]North	39,102 (53.8)	37,093 (57.7)	76,195 (55.6)
South	33,611 (46.2)	27,140 (42.3)	60,751 (44.4)
[c]Stroke Belt	11,960 (16.4)	11,689 (18.2)	23,649 (17.3)
Non-stroke Belt	60,753 (83.6)	52,544 (81.8)	11,3297(82.7)
Socio- economic region			
Low- income	20,395 (28.0)	11,909 (18.5)	32,304 (23.6)
Middle- income	23,853 (32.8)	17,097 (26.6)	40,950 (29.9)
High-income	28,464 (39.1)	35,223 (54.8)	63,691 (46.5)

Han Nationality	70,333 (96.7)	62,106 (96.7)	13,2439 (96.7)
BMI[a]			
Median (IQR)	25.0 (22.8–27.6)	25.3 (23.2–27.4)	25.2 (23.0–27.5)
Underweight (<18.5 kg/m2	1050 (1.4)	802 (1.2)	1852 (1.4)
Normal (18.5 - <24.0)	25,864 (35.6)	21,237 (33.1)	47,101 (34.4)
Overweight (24.0 - <28)	30,690 (42.2)	29,316 (45.6)	60,006 (43.8)
Obesity (≥28.0)	15,108 (20.8)	12,879 (20.1)	27,987 (20.4)
Waist circumference[†]			
Median (IQR)	86.0 (80.0–93.0)	86.0 (80.0–93.0)	86.0 (80.0–93.0)
Normal <90/85 cm (M/W)	31,077 (42.7)	25,676 (40.9)	56,753 (41.9)
Central obese ≥90/85 cm (M/W)	41,621 (57.3)	37,083 (59.1)	78,704 (58.1)
Diabetes Mellitus	20,529 (28.2)	21,008 (32.7)	41,537 (30.3)
Hypertension	46,155 (63.5)	39,280 (61.2)	85,435 (62.4)
Drinking (Current)	14,082 (19.4)	12,084 (18.8)	26,166 (19.1)
Smoking (Current)	25,677 (35.3)	19,174 (29.9)	44,851 (32.8)
Physical inactivity	23,455 (32.3)	28,599 (44.5)	52,054 (38.0)
Lipid variables: Median (IQR)			
TC mmol/l	4.92 (4.22–5.64)	4.88 (4.23–5.58)	4.90 (4.23–5.61)
HDL-C mmol/l	1.35 (1.10–1.65)	1.28 (1.08–1.55)	1.31 (1.09–1.60)
LDL-C mmol/l	2.85 (2.20–3.42)	2.85 (2.24–3.40)	2.85 (2.21–3.41)
Triglycerides mmol/l	1.50 (1.04–2.16)	1.51 (1.09–2.15)	1.50 (1.06–2.15)
Non-HDL-C [‡] mmol/l	3.51 (2.80–4.24)	3.53 (2.88–4.22)	3.52 (2.84–4.23)

Subsequent analyses documented that urban dwellers documented more lowered HDL-C levels vs. that of rural inhabitants while increased TC was more evident in the latter. This dyslipidemia documented a negative correlation with regular physical activity while the factors positively associated emerged as obesity, overweight, diabetes as well as hypertension, age and alcohol use.

Tables 5 below documents the risk factors associated with dyslipidemia in the China National Stroke Screening and prevention project (CNSSPP) participants as depicted by the researchers:

TOP: CVD factors: *Key: LDL-C* raised low-density lipoprotein; *HDL-C* low high-density lipoprotein; *TG* raised triglycerides *TC* raised total cholesterol; *N* total number of subjects (Opoku, S., Gan, Y., Fu, W. et al,

2019: *Open Access: This article is distributed under the terms of the Creative Commons Attribution 4.0 International License (http://creativecommons.org/licenses/by/4.0/), which permits unrestricted use, distribution, and reproduction in any medium, provided you give appropriate credit to the original author(s) and the source, provide a link to the Creative Commons license, and indicate if changes were made. The Creative Commons Public Domain Dedication waiver (http://creativecommons.org/publicdomain/zero/1.0/) applies to the data made available in this article, unless otherwise stated*):

	TC	HDL-C	LDL-C	TG	Non-HDL-C
Risk factors (N)	(%)	(%)	(%)	(%)	(%)
Male (70,563)	14	24	7	23	9
Female (66,383)	9	16	9	22	12
Normal/underweight (48,953)	10	17	7	16	9
Overweight (60,006)	11	21	8	24	11
Obese (27,986)	13	23	10	30	13
Diabetic (95,407)	13	22	9	26	12
Non- diabetic (41,537)	11	19	8	21	10
Hypertensive (85,435)	13	20	9	25	12
Non-hypertensive (51,510)	10	20	7	19	9
Smoker (44,851)	10	23	8	23	9
Non-Smoker (92,095)	12	19	8	22	11
Drinker (26,166)	11	20	8	26	10
Non-drinker (11,0778)	11	20	8	22	11
Physical activity (52,054)	12	20	9	24	11
Physical inactivity (84,891)	11	20	8	22	10

BOTTOM: Socioeconomic factors (Opoku, S., Gan, Y., Fu, W. et al, 2019: *Open Access: This article is distributed under the terms of the Creative Commons Attribution 4.0 International License (http://creativecommons.org/licenses/by/4.0/), which permits unrestricted use, distribution, and reproduction in any medium, provided you give appropriate credit to the original author(s) and the source, provide a link to the Creative Commons license, and indicate if changes were made. The Creative Commons Public Domain Dedication waiver (http://creativecommons.org/publicdomain/zero/1.0/) applies to the data made available in this article, unless otherwise stated*):

Key: p <0.001 except for: ^{C}p <0.05; NS, not significant (p >0.05); CNSSPP database, *HDL-C* low high-density lipoprotein; *LDL-C* raised low-density lipoprotein; *TG* raised triglycerides, *TC* raised total cholesterol; *Non- HDL-C* raised non-high-density lipoprotein

	TC	HDL-C	LDL-C	TG	Non-HDL-C
Age (years)					
40–49	1.00	1.00	1.00	1.00	1.00
50–59	1.48 (1.41–1.55)	0.86 (0.83–0.89)	1.42 (1.35–1.50)	0.95 (0.92–0.98)	1.41 (1.34–1.47)
60–69	1.70 (1.61–1.80)	0.85 (0.81 - -0.88)	1.53 (1.44–1.63)	0.84 (0.81–0.87)	1.56 (1.47–1.64)
70 and above	1.64 (1.55–1.75)	0.77 (0.74–0.81)	1.49 (1.39–1.60)	0.68 (0.65–0.71)	1.49 (1.40–1.59)
Residence					
Rural	1.00	1.00	1.00	1.00	1.00
Urban	0.90 90.87–0.93)	1.04 (1.01–1.07)[C]	0.91 (0.87–0.94)	1.06 (1.04–1.09)	0.90 (0.87–0.94)
Sex					
Men	1.00	1.00	1.00	1.00	1.00
Women	1.83 (1.75–1.91)	0.62 (0.60–0.64)	1.55 (1.47–1.63)	1.02 (1.00 - 1.06) NS	1.52 (1.45–1.59)
Socio- economic regions					
Low-income	1.00	1.00	1.00	1.00	1.00
Middle-income	1.10 (1.05–1.16)	0.85 (0.82–0.88)	0.88 (0.83–0.93)	1.35 (1.30–1.40)	1.02 (0.98–1.08) NS
High-income	1.06 (1.01–1.11)[C]	0.82 (0.79–0.85)	0.90 (0.85–0.95)	0.92 (0.88–0.95)	0.97 (0.93–1.03) NS
Ethnicities					
Han nationality	1.76 (1.56–1.99)	0.71 (0.67–0.76)	0.88 (0.79–0.98)[C]	1.14 (1.06–1.22)[C]	1.50 (1.34–1.68)
Others	1.00	1.00	1.00	1.00	1.00
BMI					
Underweight	0.93 (0.78–1.09) NS	0.87 (0.76–0.99)[C]	0.87 (0.71–1.06) NS	0.65 (0.56–0.75)	0.84 (0.71–1.01) NS
Normal weight	1.00	1.00	1.00	1.00	1.00
Overweight	1.11 (1.07–1.16)	1.15 (1.12–1.19)	1.18 (1.12–1.24)	1.41 (1.37–1.46)	1.21 (1.16–1.26)
Obesity	1.27 (1.20–1.33)	1.27 (1.22–1.33)	1.34 (1.26–1.42)	1.79 (1.72–1.87)	1.41 (1.33–1.48)
Central obesity					
Yes	1.05 (1.01–1.09)	1.24 (1,20–1 28)	1.13 (1.08–1.18)	1.25 (1.22–1.29)	1.08 (1.04–1.13)
No	1.00	1.00	1.00	1.00	1.00
History of CHD					
Yes	1.00 (0.94–1.06) NS	0.63 (0.60–0.66)	0.89 (0.83–0.96)[C]	0.77 (0,74–0.80)	0.91 (0.85–0.96)
No	1.00	1.00	1.00	1.00	1.00

History of Dyslipidemia					
Yes	1.06 (1.02–1.10)	1.09 (1.06–1.13)	1.18 (1.13–1.23)	0.98 (0.96–1.01) NS	1.09 (1.05–1.13)
No	1.00	1.00	1.00	1.00	1.00
History of Stroke					
Yes	0.92 (0.88–0.96)	0.93 (0.90–0.96)	1.00 (0.95–1.05)	0.73 (0.70–0.75)	0.91 (0.87–0.95)
No	1.00	1.00	1.00	1.00	1.00
Stroke Belt					
Yes	1.00 (0.95–1.05)	1.12 (1.07–1.17)	0.92 (0.86–0.97)[C]	1.44 (1.38–1.50)	0.99 (0.94–1.05) NS
No	1.00	1.00	1.00	1.00	1.00
Geographical region					
North	1.12 (1.07–1.17)	0.61(0.59–0.63)	1.13 (1.08–1.19)	0.67 (0.65–0.69)	0.98 (0.94–1.02) NS
South	1.00	1.00	1.00	1.00	1.00
Diabetes Mellitus					
Yes	1.14 (1.10–1.19)	1.18 (1.14–1.21)	1.17 (1.12–1.22)	1.25 (1.22–1.29)	1.21 (1.16–1.25)
No	1.00	1.00	1.00	1.00	1.00
Hypertension					
Yes	1.20 (1.15–1.24)	0.95 (0.92–0.97)	1.23 (1.18–1.29)	1.23 (1.19–1.26)	1.20 (1.15–1.24)
No	1.00	1.00	1.00	1.00	1.00
Current smoker					
Yes	1.02 (0.97–1.07) NS	1.01 (0.97–1.04) NS	1.21 (1.15–1.28)	0.99 (0.96–1.03) NS	1.06 (1.01–1.12) C
No	1.00	1.00	1.00	1.00	1.00
Current drinker					
Yes	1.33 (1.27–1.40)	0.76 (0.73–0.79)	1.15 (1.08–1.22)	1.18 (1.13–1.22)	1.16 (1.10–1.22)
No	1.00	1.00	1.00	1.00	1.00
Physical activity					
Yes	0.96 (0.92–0.99)[C]	0.95 (0.92–0.97)	0.94 (0.90–0.98)[C]	0.89 (0.86–0.91)	0.96 (0.92–0.99) C
No	1.00	1.00	1.00	1.00	1.00

2.5 A GLIMPSE FROM LATIN AMERICA AND THE CARIBBEAN

Carrillo-Larco and team (2020) reported a review and meta-analysis of population-based studies of the profile of mean total cholesterol and lipid fractions with dyslipidaemias in Latin America and the Caribbean (LAC). Studies between 1964 and 2016 entailing "*(total cholesterol [TC; 86 studies; 168,553 people], HDL-Cholesterol [HDL-C; 84 studies; 121,282 people],*

LDL-Cholesterol [LDL-C; 61 studies; 86,854 people], and triglycerides [TG; 84 studies; 121,009 people])" revealed no evident change in all four lipid parameters scrutinized for the time span that is indicative of either ineffective implementation or paucity of control measures in the areas assessed.

Figure 6 below illustrates the trends in mean total cholesterol, LDL-Cholesterol, HDL-Cholesterol and triglycerides in Latin America and the Caribbean with the year of data collection on the x-axis and the regression trend in blue as depicted by the authors (eLife 2020;9:e57980 DOI: 10.7554/eLife.57980 © 2020, Carrillo-Larco et al: *This article is distributed under the terms of the Creative Commons Attribution License, which permits unrestricted use and redistribution provided that the original author and source are credited*):

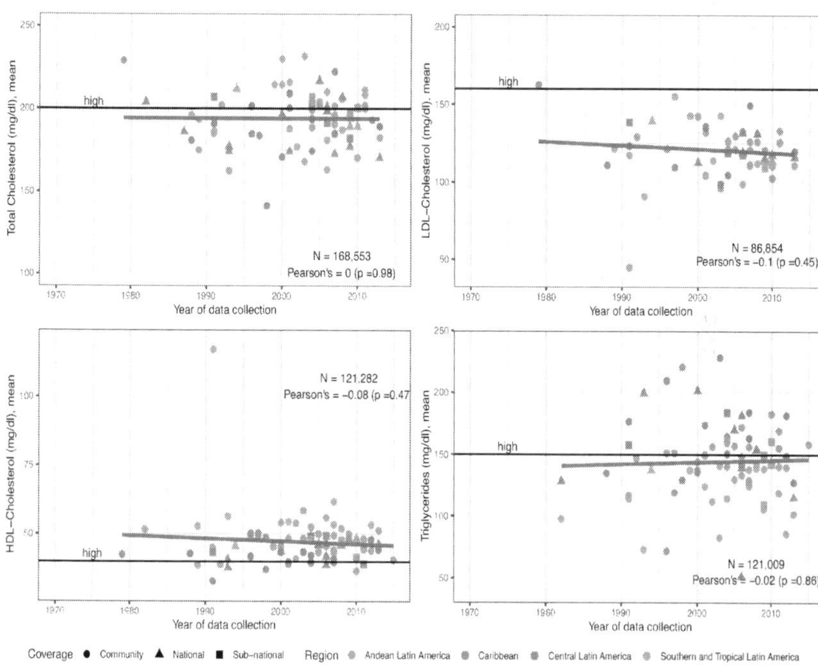

2.6 CARDIOVASCULAR RISK PREVENTION (CRISP) IN ASIA NETWORK

Lee and team (2019) reviewed national surveys of aberrant lipid profiles as conducted by government agencies in six countries: Indonesia, Malaysia, Philippines, Thailand, and Vietnam with Australia serving as the reference.

Philippines (*2015 Clinical Practice Guidelines for the Management of Dyslipidemia in the Philippines*) [43]	• Risk factor counting to identify patients in need of statins	• <130 mg/dL (or a 30% reduction) for those at lower risk • <70 mg/dL (or > 30% reduction) for those with established ASCVD	• Statins for non-diabetic patients aged ≥45 years with LDL-C ≥ 130 mg/dL and ≥ 2 risk factors without atherosclerotic CVD and for diabetic individuals without atherosclerotic CVD	Regardless of age of CV risk, guidelines recommend initiation of moderate- intensity statin therapy	• High-intensity statin (based on LDL-C reduction)
Thailand (*2016 RCPT Clinical Practice Guideline on Pharmacologic Therapy of Dyslipidemia for Atherosclerotic Cardiovascular Disease Prevention*) [44]	• CV risk score used based on Thai patient population (Thai CV Risk Score)	• <130 mg/dL (or a 30% reduction) for those with 10-year risk ≥10% • <100 mg/dL for primary prevention in DM, CKD, or familial hypercholesterolemia • <70 mg/dL (or a 50% reduction) for clinical ASCVD	• Statins for patients with LDL-C ≥ 190 mg/dL, familial hypercholesterolemia, 10-year risk ≥10%	• Statins for patients: - DM ≥40 years - DM <40 years + 2 CV risk factors + LDL-C ≥ 100 mg/dL (moderate intensity statins) - DM <40 years with 0 or 1 CV risk factor + LDL-C ≥ 100 mg/dl (low to moderate intensity statins)	• Statins (moderate to high intensity depending on atherosclerotic CVD) • Non-statin if LDL-C target not reached in 6 months
Vietnam (*Recommendations on Diagnosis and Treatment of Lipid Disorders 2015*)	• SCORE scale (low risk) used • Stratification of CV risk as low, medium, high, or very high	ESC recommendations for goals Primary < 100 mg/dL Secondary <70 mg/dL	Statins for patients with LDL-C ≥ 190 mg/dL	• Statins for patients aged 40–75 with diabetes and LDL-C between 70 and 189 mg/dL • High-intensity statins used for most patients with diabetes	• Statins for patients with atherosclerosis, acute coronary syndrome, history of MI, stable/unstable angina • LDL < 70 mg/dL

2.7 REPOSITIONING GLOBAL PATTERNS IN CHOLESTEROL

2020-published research by the NCD Risk Factor Collaboration (NCD-RisC) in *Nature* employed a Bayesian hierarchical model to scrutinize blood lipids viz. mean total, non-HDL and HDL cholesterol levels in 102.6 million individuals (above 18 years of age) of 200 countries from 1,127 population-based studies from 1980 to 2018.

The global profile of ***total cholesterol*** did not document any significant change: 0.03 mmol l^{-1} per decade (−0.02–0.08) in women and 0.05 mmol l^{-1} per decade (0.00–0.11) in men in the time span probed. High-income western regions and in central and eastern Europe, notably northwestern Europe documented a conspicuous decrease in total cholesterol. East and Southeast Asia displayed an increase of more than 0.1 mmol l^{-1} per decade in both genders due to an elevation in non-HDL cholesterol.

Figure 7 below illustrates the profile of age-standardized mean total cholesterol in **a)** women; **b)** men. The arrowhead is for 2018 and the start is that for 1980 (NCD Risk Factor Collaboration (NCD-RisC), 2020: *Open Access: This article is licensed under a Creative Commons Attribution 4.0 International License, which permits use, sharing, adaptation, distribution and reproduction in any medium or format, as long as you give appropriate credit to the original author(s) and the source, provide a link to the Creative Commons license, and indicate if changes were made. The images or other third party material in this article are included in the article's Creative Commons license, unless indicated otherwise in a credit line to the material. If material is not included in the article's Creative Commons license and your intended use is not permitted by statutory regulation or exceeds the permitted use, you will need to obtain permission directly from the copyright holder. To view a copy of this license, visit http://creativecommons.org/licenses/by/4.0/*):

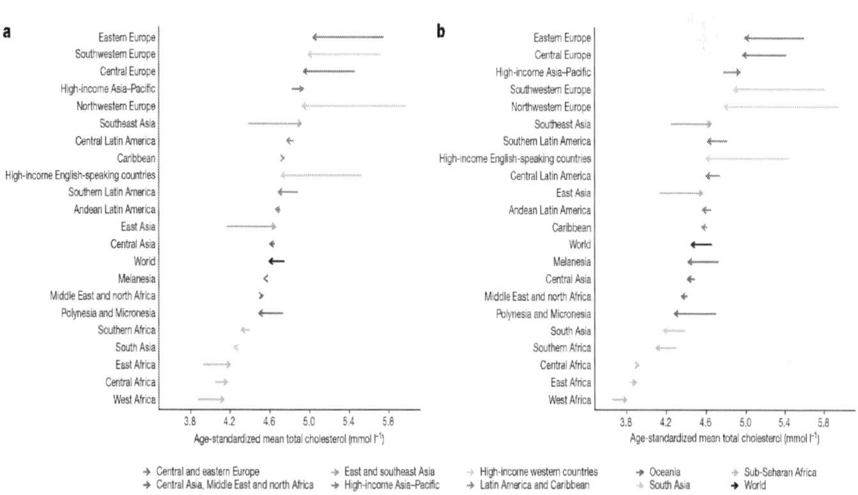

The global ***non-HDL levels*** also remained relatively unchanged between 1980 and 2018 while locally high-income western regions and central and eastern Europe, notably northwestern Europe documenting an evident decrease. East and Southeast Asia, sections of sub-Saharan Africa and Melanesia documented increases. Figure 8 below illustrates the profile of age-standardized mean non-HDL cholesterol in **a)** women; **b)** men. The arrowhead is for 2018 and the start is that for 1980 (NCD Risk Factor Collaboration (NCD-RisC), 2020: *Open Access: This article is licensed under a Creative Commons Attribution 4.0 International License, which permits use, sharing, adaptation, distribution and reproduction in any medium or format, as long as you give appropriate credit to the original author(s) and the source, provide a link to the Creative Commons license, and indicate if changes were made. The images or other third party material in this article are included in the article's Creative Commons license, unless indicated otherwise in a credit line to the material. If material is not included in the article's Creative Commons license and your intended use is not permitted by statutory regulation or exceeds the permitted use, you will need to obtain permission directly from the copyright holder. To view a copy of this license, visit http://creativecommons.org/licenses/by/4.0/*):

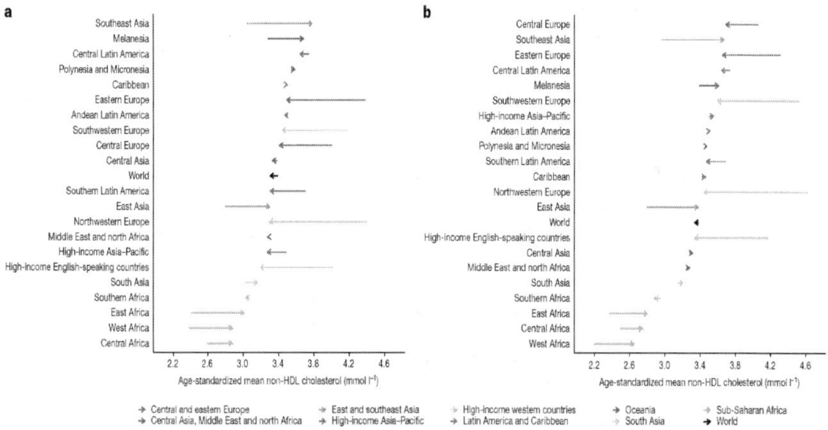

The countries with elevated non-HDL in 1980 inclusive of Belgium, Finland, Greenland, Iceland, Norway, Sweden, Switzerland and Malta caused a shift in their ranking on the basis of non-HDL cholesterol from being in the top 10 countries in 1980 to lower ranks in 2018. The opposite trends emerged for east and southeast Asia: China documented the lowest non-HDL cholesterol levels in 1980 touched the ranks of high-income western countries in 2018. Overall, a shift emerged in the levels of non-HDL cholesterol—a cardiovascular risk marker from 1980 to 2018 i.e. from "*Belgium, Finland, Greenland, Iceland, Norway, Sweden, Switzerland and Malta in 1980 to those in Asia and the Pacific, such as Tokelau, Malaysia, The Philippines and Thailand*".

Figure 9 below illustrates ***TOP:*** the age-standardized mean non-HDL cholesterol profile in **a)** women; 1980; **b)** women, 2018; **c)** men, 1980 and **d)** mean; 2018 (NCD Risk Factor Collaboration (NCD-RisC), 2020: *Open Access: This article is licensed under a Creative Commons Attribution 4.0 International License, which permits use, sharing, adaptation, distribution and reproduction in any medium or format, as long as you give appropriate credit to the original author(s) and the source, provide a link to the Creative Commons license, and indicate if changes were made. The images or other third party material in this article are included in the article's Creative Commons license, unless indicated otherwise in a credit line to the material. If material is not included in the article's Creative Commons license and your intended use is not permitted by statutory regulation or exceeds the permitted use, you will need to obtain permission directly from the copyright holder. To view a copy of this license, visit http://creativecommons.org/licenses/by/4.0/*):

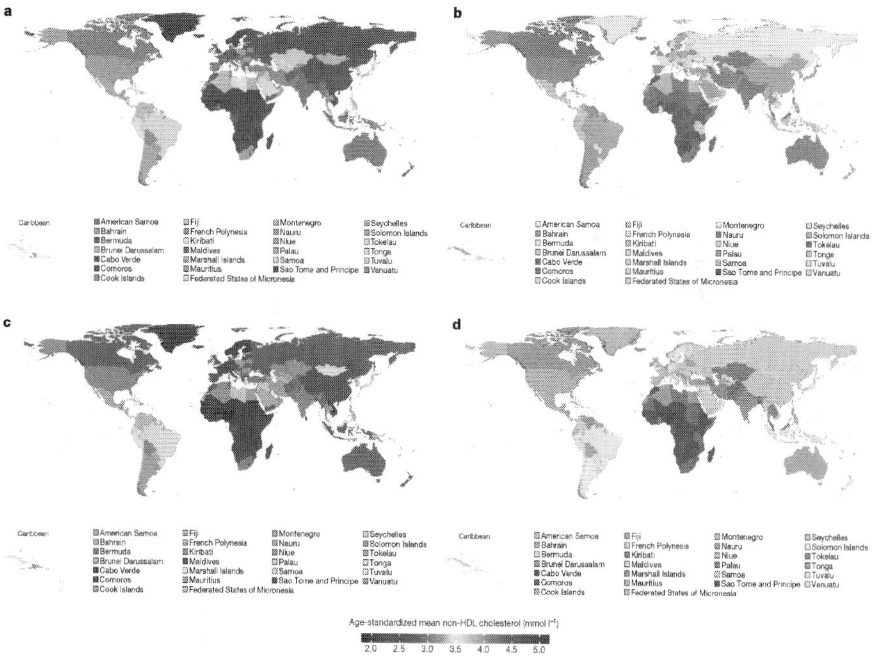

BOTTOM: Alteration in age-standardized mean non-HDL cholesterol per decade for **a)** women; **b)** men (NCD Risk Factor Collaboration (NCD-RisC), 2020: *Open Access: This article is licensed under a Creative Commons Attribution 4.0 International License, which permits use, sharing, adaptation, distribution and reproduction in any medium or format, as long as you give appropriate credit to the original author(s) and the source,*

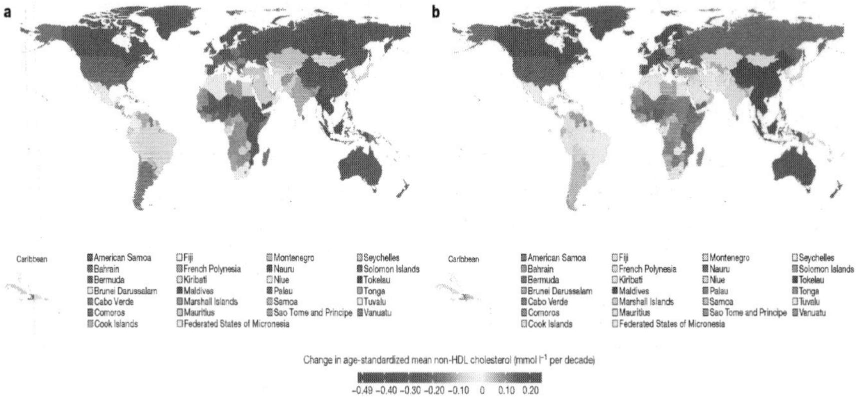

Globally, IHD and ischaemic stroke accounted for elevated non-HDL cholesterol that was attributed to close to 3.9 million (3.7–4.2 million) mortalities in 2017. While these numbers were lowered from 950,000 (890,000–990,000) to 480,000 (430,000–530,000) in western countries, Asia documented an increase from 110,000 (100,000–120,000) to 310,000 (290,000–330,000) in southeast Asia and from 250,000 (230,000–270,000) to 860,000 (770,000–940,000) in east Asia. Such a shift in the cholesterol profile warrants "universal health coverage" encompassing treatment regimens targeting specific populations.

"There's a way around almost every obstacle, you just need to find it" -Julie Randall"

2.8 REFERENCES

1. NCD Risk Factor Collaboration (NCD-RisC) (2020). National trends in total cholesterol obscure heterogeneous changes in HDL and non-HDL cholesterol and total-to-HDL cholesterol ratio: a pooled analysis of 458 population-based studies in Asian and Western countries. *International journal of epidemiology*, *49*(1), 173–192.https://doi.org/10.1093/ije/dyz099
2. Gupta, R., Guptha, S., Agrawal, A. et al. Secular trends in cholesterol lipoproteins and triglycerides and prevalence of dyslipidemias in an urban Indian population. Lipids Health Dis 7, 40 (2008). https://doi.org/10.1186/1476-511X-7-40
3. Joshi SR, Anjana RM, Deepa M, Pradeepa R, Bhansali A, Dhandania VK, et al. (2014) Prevalence of Dyslipidemia in Urban and Rural India: The ICMR–INDIAB Study. PLoS ONE 9(5): e96808. https://doi.org/10.1371/journal.pone.0096808
4. Kaufman HW, Blatt AJ, Huang X, Odeh MA, SuperkoHR (2013) Blood Cholesterol Trends 2001–2011 in the United States: Analysis of 105 Million Patient Records. PLoS ONE 8(5): e63416. https://doi.org/10.1371/journal.pone.0063416
5. Wang S, Xu L, Jonas JB, You QS, Wang YX, Yang H. Prevalence and associated factors of dyslipidemia in the adult Chinese population. PLoS One. 2011 Mar 10;6(3):e17326. doi: 10.1371/journal.pone.0017326. PMID: 21423741; PMCID: PMC3053360.
6. Aranmolate, R. and Obayemi, O. S. (2018). Trends in the Prevalence and Awareness of High Cholesterol among Adults in Mississippi, United States. European Journal of Environment and Public Health, 2(2), 06. https://doi.org/10.20897/ejeph/92010
7. Opoku, S., Gan, Y., Fu, W. et al. Prevalence and risk factors for dyslipidemia among adults in rural and urban China: findings from the China National Stroke Screening and prevention project (CNSSPP). BMC Public Health 19, 1500 (2019). https://doi.org/10.1186/s12889-019-7827-5
8. Carrillo-Larco et al. A systematic review of population-based studies on lipid profiles in Latin America and the Caribbean. eLife 2020;9:e57980. DOI: https://doi.org/10.7554/eLife.57980

9. Lee, ZV., Llanes, E.J., Sukmawan, R. et al. Prevalence of plasma lipid disorders with an emphasis on LDL cholesterol in selected countries in the Asia-Pacific region. Lipids Health Dis 20, 33 (2021). https://doi.org/10.1186/s12944-021-01450-8
10. NCD Risk Factor Collaboration (NCD-RisC). Repositioning of the global epicentre of non-optimal cholesterol. Nature 582, 73–77 (2020). https://doi.org/10.1038/s41586-020-2338-1

CHAPTER

3

Cholesterol And Covid-19 Research Insights

Contents

3.1 GROUNDWORK

"It has been shown that high cholesterol levels are associated with more lipid rafts, subdomains of the plasma membrane that can harbour angiotensin-converting enzyme 2 (ACE2) receptors for the S-protein of SARS-CoV-2" -Radenkovic, D., Chawla, S., Pirro, M., Sahebkar, A., & Banach, M. (2020). Cholesterol in Relation to COVID-19: Should We Care about It?. Journal of clinical medicine, 9(6), 1909. https://doi.org/10.3390/jcm9061909

This chapter looks at studies probing the cholesterol profile (with other relevant molecules) in COVID-19 patients. This becomes pertinent given the multipronged impact of the Severe Acute Respiratory Syndrome Coronavirus (SARS-CoV-2), the causal agent of Coronavirus Disease 2019 (COVID-19) pandemic. Categorized on the basis of density (low to high density), the composition of lipids and proteins define lipoprotein particles that are also linked to a slew of functions apart from cholesterol transport. Early work in 2008 probed the membrane alterations induced by coronaviruses employing electron tomography of infected cells to reveal "a unique reticulovesicular network of modified endoplasmic reticulum that integrates convoluted membranes, numerous interconnected double-membrane vesicles or DMVs (diameter 200–300 nm), and "vesicle packets". This membrane reorganization was purported to facilitate viral replication and viral RNA evasion (Knoops et al, 2008). Figure 1 below illustrates the reticulovesicular network of modified membranes induced by SARS-CoV documented by the researchers: double-stranded RNA (dsRNA), double-membrane vesicle (DMVs), nonstructural protein (nsp) and vesicle packets (VPs) (© 2008 Knoops et al. *This is an open-access article distributed under the terms of the Creative Commons Attribution License, which permits unrestricted use, distribution, and reproduction in any medium, provided the original author and source are credited*):

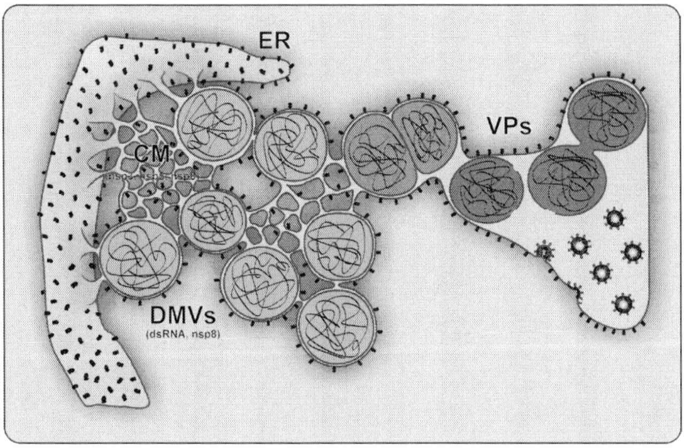

In an interesting work titled "Building Viral Replication Organelles: Close Encounters of the Membrane Types" by Nagy, Strating and van Kuppeveld (2016), the involvement of virus-induced membrane contact sites (vMCSs) in channelizing lipids by +-strand RNA viruses inclusive of SARS-coronavirus, Zika virus and Dengue virus. This has been depicted in figure 2 below (Nagy, Strating and van Kuppeveld © 2016 Nagy et al. *This is an open access article distributed under the terms of the Creative Commons Attribution License, which permits unrestricted use, distribution, and reproduction in any medium, provided the original author and source are credited*): A) Interaction of viral proteins with VAMP-associated protein (VAP) and OSBP-related proteins or ORPs facilitates sterol enrichment in the peroxisomal membrane. B) Viral replicase complex (VRC) and vMCS in an infected plant cell shown by electron microscopy. C) Recruitment of oxysterol binding protein (OSBP) that facilitates cholesterol accumulation in membranes of organelles.

In this context, HDLs and LDLs have been documented to bind, neutralize and clear lipopolysaccharides (LPS). Further, sepsis has been linked to lowered HDL-C and LDL levels that is further also connected to poor outcomes. Scrutiny in animal models has unveiled the protective impact of reconstituted HDL (rHDL) and the ApoA-I mimetic peptide in septic shock (Tanaka et al, 2020).

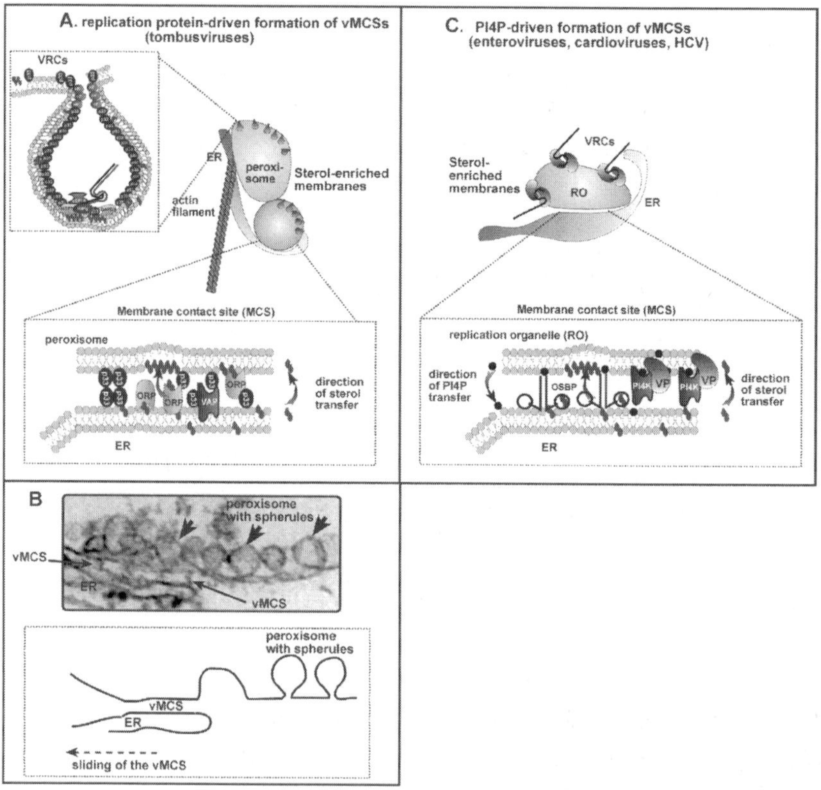

Researchers Dias and team (2020) documented an increased accumulation of lipid droplets (LD) vitally involved in transport, energy balance and inflammation in monocytes from COVID-19 patients as opposed to individuals without a SARS-CoV-2 infection. Proteins associated with lipid metabolism were found to be upregulated in human primary monocytes infected with SARS-CoV-2 inclusive of CD36, PPARγ, SREBP-1, and DGAT-1. Inhibiting the enzyme DGAT-1 with A922500 quelled the LD formation triggered by SARS-CoV-2 infection in A549 human epithelial cells and in primary human monocytes in a dose-dependent manner. The treatment also diminished the viral load in human primary monocytes in a dose-dependent way.

Figure 3 below outlines the findings of Dias and team (2020): ***TOP***: (A) Representative images of monocytes from COVID-19 patients and healthy volunteers, Oil Red O staining (Red) and DAPI-stained nuclei (Blue). (C) Representative images of SARS-CoV-2 infected human monocytes isolated from PBMCs Scale bar 20μm (B and D) Fluorescence analyses of (E) SARS-CoV-2 and lipid metabolism (F) CD36, PPAR-γ, SREBP-1, DGAT-1 in SARS-CoV-2-infected monocytes with β-as control (G) Densitometric analyses of 1F (© 2020 Dias et al. *This is an open access article distributed under the terms of the Creative Commons Attribution License, which permits unrestricted use, distribution, and reproduction in any medium, provided the original author and source are credited*):

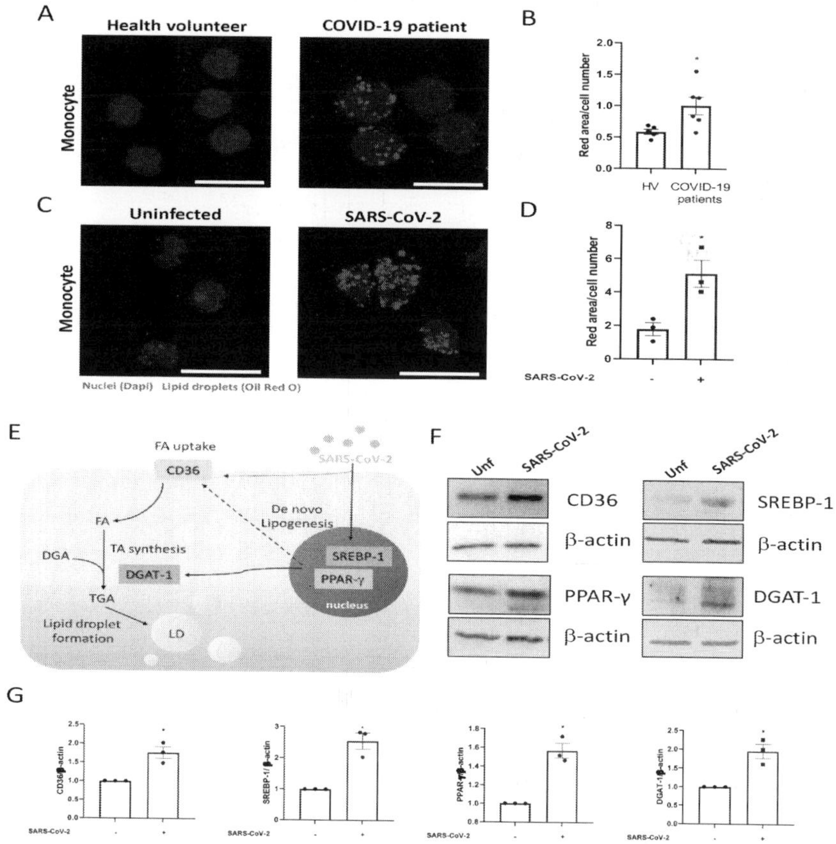

BOTTOM: The lowered pro-inflammatory profile, cell death and viral load induced by SARS-CoV-2 in monocytes treated with the DGAT-1 inhibitor A922500 (A) Cell death (B) Viral load by qPCR (C) Phase-contrast images of monocytes. Scale bar 20μm. (D-F) Measuring inflammatory cytokines (D) leukotrienes: CysLT and LTB4, (E) chemokines: IL-8 and CXCL10, (F) inflammatory cytokines: IL-6, TNF-α and IL-10 (© 2020 Dias et al. *This is an open access article distributed under the terms of the Creative Commons Attribution License, which permits unrestricted use, distribution, and reproduction in any medium, provided the original author and source are credited*):

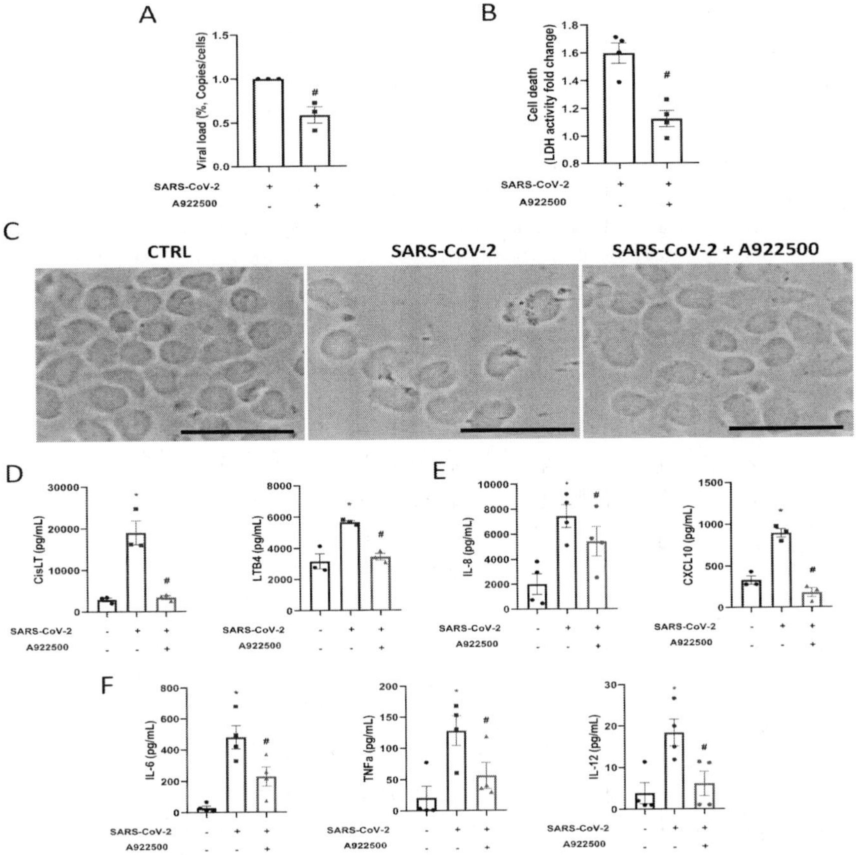

Research by Wang and team (2021) in *Cell* reported that phosphatidylinositol phosphate biosynthesis and cholesterol homeostasis were vital host factors supporting the infection of SARS-CoV-2 along with OC43 and 229E: two seasonally circulating common cold coronaviruses employing genome-wide CRISPR screens.

This now sets the path for further probing of the involvement of cholesterol and associated lipids in COVID-19.

3.2 A GLOBAL PICTURE

Researchers Sarmadi and team (2021) probed the association between Body Mass Index (BMI) and cholesterol levels with COVID-19 incidence and mortality. While COVID-19 data were sourced from the World Health Organization website (WHO Coronavirus Disease Dashboard, https://covid19.who.int/), the WHO and Global Burden Of Diseases (GBD) and Non-Communicable Diseases Risk Factor Collaboration (NCD-RisC) served as data sources for BMI and cholesterol levels. Analyses unveiled stronger correlations of BMI and cholesterol level with COVID-19 in countries having a younger population.

Figure 4 below illustrates the GIS distribution map of; a COVID-19case ratio b COVID-19 death ratio, c mean age-standardized BMI, d % overweight prevalence (age-standardized), e % obesity prevalence (age-standardized), f mean cholesterol level (age-standardized), g % increased total cholesterol ≥ 5 mmol/L (age-standardized) and h) % increased total cholesterol ≥ 6.2 mmol/L (age-standardized) (Sarmadi, M., Ahmadi-Soleimani, S.M., Fararouei, M. et al, 2021: *Open Access: This article is licensed under a Creative Commons Attribution 4.0 International License, which permits use, sharing, adaptation, distribution and reproduction in any medium or format, as long as you give appropriate credit to the original author(s) and the source, provide a link to the Creative Commons licence, and indicate if changes were made. The images or other third party material in this article are included in the article's Creative Commons licence, unless indicated otherwise in a credit line to the material. If material is not included in the article's Creative Commons licence and your intended use is not permitted by statutory regulation or exceeds the permitted use, you will need to obtain permission directly from the copyright holder. To view a copy of this licence, visit http://creativecommons.org/licenses/by/4.0/. The Creative Commons Public Domain Dedication waiver (http://creativecommons.org/publicdomain/zero/1.0/) applies to the data made available in this article, unless otherwise stated in a credit line to the data*):

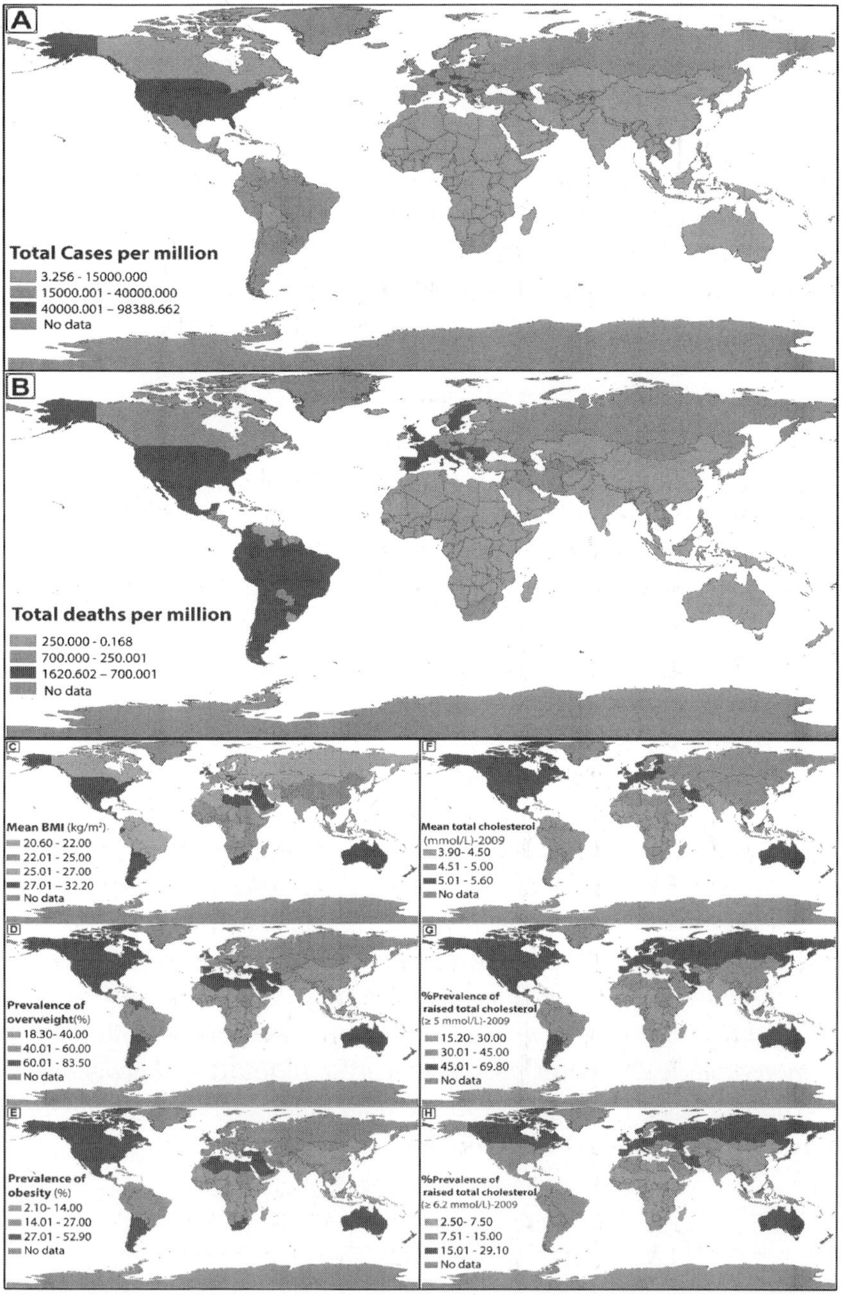

Following socioeconomic and demographic factor adjustment, a positive correlation emerged for the BMI and cholesterol level with COVID-19 incidence ratio ($\beta = 2396.81$ and $\beta = 30{,}932.80$, $p < 0.01$, respectively) and mortality ratio ($\beta = 38.18$ and $\beta = 417.52$, $p < 0.05$, respectively). Table

1 below documents the correlation values as presented by the researchers (Sarmadi, M., Ahmadi-Soleimani, S.M., Fararouei, M. et al, 2021: *Open Access: This article is licensed under a Creative Commons Attribution 4.0 International License, which permits use, sharing, adaptation, distribution and reproduction in any medium or format, as long as you give appropriate credit to the original author(s) and the source, provide a link to the Creative Commons licence, and indicate if changes were made. The images or other third party material in this article are included in the article's Creative Commons licence, unless indicated otherwise in a credit line to the material. If material is not included in the article's Creative Commons licence and your intended use is not permitted by statutory regulation or exceeds the permitted use, you will need to obtain permission directly from the copyright holder. To view a copy of this licence, visit http://creativecommons.org/licenses/by/4.0/. The Creative Commons Public Domain Dedication waiver (http://creativecommons.org/publicdomain/zero/1.0/) applies to the data made available in this article, unless otherwise stated in a credit line to the data*):

**GDP Gross Domestic Product per capita, ** P < .01, * P < .05, a = per million*

Variables	Total cases	Total deaths	Incidence ratio [a]	Mortality ratio [a]
Mean total serum cholesterol (2009)	.495**	.391**	.692**	.611**
Prevalence of high cholesterol (≥ 5)	.498**	.397**	.695**	.617**
Prevalence of high cholesterol (≥ 6.2)	.485**	.383**	.701**	.621**
Mean total cholesterol (mmol/L)	.450**	.409**	.609**	.602**
Mean non-HDL cholesterol (mmol/L)	.419**	.411**	.488**	.539**
Mean HDL cholesterol (mmol/L)	.199*	.139	.476**	.390**
Mean BMI (kg/m^2)	.364**	.356**	.586**	.617**
Prevalence of overweight among adults	.438**	.410**	.641**	.643**
Prevalence of obesity among adults	.388**	.374**	.588**	.604**
Prevalence of underweight among adults	−.457**	−.436**	−.704**	−.715**
GDP per capita	.475**	.356**	.679**	.564**
Median age	.523**	.447**	.679**	.621**
Aged > 65 older	.484**	.459**	.582**	.605**
Aged > 70 older	.480**	.454**	.594**	.615**
Population density	.149	.066	.141	.033

This further warrants detailed probing of the association of cholesterol and COVID-19.

3.3 THE LIPID PROFILE IN COVID-19:

The lipid profile in the blood of COVID-19 patients was probed by Wang and team (2020) in *Lipids in Health and Disease* with 228 adults (confirmed by reverse transcription-polymerase chain reaction or RT-PCR) admitted to the Public Health Treatment Center of Changsha, China, from January 17 to March 14, 2020, vs. 1140 control patients. Analyses unveiled a lowering of total cholesterol (TC), high-density lipoprotein cholesterol (HDL-C), low-density lipoprotein cholesterol (LDL-C) and glyceride in the COVID-19 patients vs. controls. Among patients, HDL-C was lowered in those with severe infections vs. non-severe infections.

Table 2 below documents the lipid profile in ***TOP***: controls vs. COVID-19 patients (Wang, G., Zhang, Q., Zhao, X. et al, 2020: *Open Access: This article is licensed under a Creative Commons Attribution 4.0 International License, which permits use, sharing, adaptation, distribution and reproduction in any medium or format, as long as you give appropriate credit to the original author(s) and the source, provide a link to the Creative Commons licence, and indicate if changes were made. The images or other third party material in this article are included in the article's Creative Commonslicence, unless indicated otherwise in a credit line to the material. If material is not included in the article's Creative Commons licence and your intended use is not permitted by statutory regulation or exceeds the permitted use, you will need to obtain permission directly from the copyright holder. To view a copy of this licence, visit http://creativecommons.org/licenses/by/4.0/. The Creative Commons Public Domain Dedication waiver (http://creativecommons.org/publicdomain/zero/1.0/) applies to the data made available in this article, unless otherwise stated in a credit line to the data*):

Bold lettering shows statistical significance at P < 0.05:

	Normal range	**Controls**	**Patients**	***P*-value**
Age, median (IQR), y		45.5 (36.0–60.8)	45.5 (36.0–60.8)	1.000
Sex (male/female)		575/565	115/113	1.000
Total cholesterol, median (IQR), mmol/L	2.33–5.69	4.65 (3.92, 5.49)	3.76 (3.22, 4.26)	**0.031**

Triglyceride, median (IQR), mmol/L	0.25–1.71	1.21 (0.93, 1.36)	1.08 (0.78, 1.44)	**< 0.001**
HDL-C, median (IQR), mmol/L	0.90–1.94	1.37 (1.22, 1.51)	0.78 (0.66, 0.97)	**< 0.001**
LDL-C, median (IQR), mmol/L	0.60–4.14	2.83 (2.27, 3.39)	2.63 (2.21, 3.09)	**< 0.001**

BOTTOM: Severe vs. non-severe infections (Wang, G., Zhang, Q., Zhao, X. et al, 2020: *Open Access: This article is licensed under a Creative Commons Attribution 4.0 International License, which permits use, sharing, adaptation, distribution and reproduction in any medium or format, as long as you give appropriate credit to the original author(s) and the source, provide a link to the Creative Commons licence, and indicate if changes were made. The images or other third party material in this article are included in the article's Creative Commonslicence, unless indicated otherwise in a credit line to the material. If material is not included in the article's Creative Commons licence and your intended use is not permitted by statutory regulation or exceeds the permitted use, you will need to obtain permission directly from the copyright holder. To view a copy of this licence, visit http:// creativecommons.org/licenses/by/4.0/. The Creative Commons Public Domain Dedication waiver (http://creativecommons.org/publicdomain/ zero/1.0/) applies to the data made available in this article, unless otherwise stated in a credit line to the data*):

Bold lettering shows statistical significance at P < 0.05:

	Normal range	**Severe (*n* = 44)**	**Non-severe (*n* = 184)**	***P-value***
Total cholesterol, median (IQR), mmol/L	2.33–5.69	3.63 (3.04–4.15)	3.81 (3.24–4.34)	0.082
Triglyceride, median (IQR), mmol/L	0.25–1.71	1.08 (0.76–1.36)	1.09 (0.79–1.47)	0.382
HDL-C, median (IQR), mmol/L	0.90–1.94	0.69 (0.59–0.95)	0.79 (0.69–0.97)	**0.032**
LDL-C, median (IQR), mmol/L	0.60–4.14	2.60 (2.19–2.95)	2.65 (2.22–3.10)	0.233

An increase in C-reactive protein (CRP) was documented in patients with the lowered HDL-C-this emerges as pertinent given that CRP is an inflammatory marker. Further modeling unveiled a higher risk of adverse outcomes in the patient with diminished HDL-C.

Table 3 below documents ***TOP***: the characters of patients: *Bold lettering shows statistical significance at P < 0.05*(Wang, G., Zhang,

	Low HDL-C (*n* = 46)	**High HDL-C (*n* = 182)**	***P*-value**
Sex (male/female)	32/14	83/99	**0.004**
Age, median (IQR), y	46.0 (37.8, 59.0)	45.0 (35.0, 61.0)	0.795
Comorbidity			
Hypertension (n, %)	9 (19.57)	27 (14.84)	0.432
Cardiovascular disease (n, %)	1 (2.17)	8 (4.40)	0.789
Diabetes (n, %)	5 (10.87)	10 (5.49)	0.327
Chronic liver disease (n, %)	5 (10.87)	7 (3.85)	0.124
Symptoms			
Fever (n, %)	38 (82.61)	135 (74.18)	0.232
Pharyngalgia (n, %)	5 (10.87)	29 (15.93)	0.389
Cough (n, %)	41 (89.13)	146 (80.22)	0.160
Expectoration (n, %)	24 (52.17)	81 (44.51)	0.351
Dyspnea (n, %)	20 (43.48)	60 (32.97)	0.182
Hemoptysis (n, %)	3 (6.52)	4 (2.20)	0.298
Chills (n, %)	5 (10.87)	24 (13.19)	0.673
Myalgia (n, %)	6 (13.04)	18 (9.89)	0.724
Fatigue (n, %)	24 (52.17)	82 (45.05)	0.387
Dizziness (n, %)	3 (6.52)	26 (14.29)	0.158
Headache (n, %)	1 (2.17)	30 (16.48)	**0.011**
Diarrhea (n, %)	11 (23.91)	41 (22.53)	0.841

Nausea (n, %)	2 (4.35)	28 (15.38)	**0.048**
Anorexia (n, %)	26 (56.52)	88 (48.35)	0.322
Vomiting (n, %)	4 (8.70)	21 (11.54)	0.581
Abdominal pain (n, %)	1 (2.17)	6 (3.30)	1.000
Chest CT positive rate (n, %)	46 (100.0)	172 (94.51)	0.221
Chest CT with ground-glass change (n, %)	20 (43.48)	88 (48.35)	0.554
Severe cases (n, %)	17 (36.96)	27 (14.84)	**0.001**
Length of hospital stay, median (IQR), days	13.0 (11.0, 22.5)	16.0 (11.0, 25.0)	0.712
Virus shedding duration, median (IQR), days	17.0 (13.5, 24.0)	18.0 (13.0, 26.0)	0.269
Mortality (n, %)	1 (2.17)	1 (0.55)	0.364

BOTTOM: Lab findings in patients *Bold lettering shows statistical significance at P < 0.05:* (Wang, G., Zhang, Q., Zhao, X. et al, 2020: *Open Access: This article is licensed under a Creative Commons Attribution 4.0 International License, which permits use, sharing, adaptation, distribution and reproduction in any medium or format, as long as you give appropriate credit to the original author(s) and the source, provide a link to the Creative Commons licence, and indicate if changes were made. The images or other third party material in this article are included in the article's Creative Commonslicence, unless indicated otherwise in a credit line to the material. If material is not included in the article's Creative Commons licence and your intended use is not permitted by statutory regulation or exceeds the permitted use, you will need to obtain permission directly from the copyright holder. To view a copy of this licence, visit http://creativecommons.org/licenses/by/4.0/. The Creative Commons Public Domain Dedication waiver (http://creativecommons.org/publicdomain/zero/1.0/) applies to the data made available in this article, unless otherwise stated in a credit line to the data*):

	Normal range	**Low HDL (*n* = 46)**	**High HDL (*n* = 182)**	***P*-value**
White blood cell count, $\times 10^9$/L, median (IQR)	4–10	4.78 (3.82, 6.13)	4.58 (3.52, 5.66)	0.627
Lymphocyte count, $\times 10^9$/L, median (IQR)	0.8–4.0	1.02 (0.72, 1.51)	1.17 (0.84, 1.59)	0.207
Lymphocyte %, median (IQR)	20–40	22.35 (17.58, 30.10)	27.45 (20.05, 33.00)	0.050
Alanine aminotransferase, U/L, median (IQR)	0–42	21.49 (16.13, 30.28)	18.81 (13.81, 26.61)	**0.044**
Aspartate aminotransferase, U/L, median (IQR)	0–37	25.87 (19.53, 34.01)	23.76 (19.25, 28.77)	0.109
Total bilirubin, μmol/L, median (IQR)	3.4–20.5	11.16 (8.16, 15.35)	10.87 (8.68, 16.36)	0.769
C-reactive protein, mg/L, median (IQR)	0–8	27.83 (14.68, 44.74)	12.56 (3.29, 26.88)	**< 0.001**
Erythrocyte sedimentation rate, mm/h, median (IQR)	0–15	52.00 (22.00, 67.75)	39.50 (22.0, 68.25)	0.538
Procalcitonin, ≥0.05 ng/mL, No. (%)	< 0.05	17 (37.0%)	46 (25.3%)	0.113
Creatinine, μmol/L, median (IQR)	21.5–104	53.70 (43.61, 65.02)	50.85 (41.22, 64.02)	0.209
Creatine kinase, U/L, median (IQR)	10–190	83.35 (57.05, 148.63)	69.75 (46.30, 114.45)	0.116
Creatine kinase-MB, U/L, median (IQR)	0–24	9.35 (5.33, 12.50)	9.57 (6.40, 12.92)	0.624

Researchers Tanaka and team (2020) probed the lipoprotein profile in patients with severe COVID-19 in the surgical ICU of Bichat Claude-Bernard University Hospital, Paris, France (ApoCOVID study, IRB 00010254-2020-082). The study entailed 48 ICU patients admitted from March 18, 2020, to April 13, 2020. The authors reported a lowering of the lipoprotein levels than reference values: ([HDL-C] = 0.7[0.5–0.9] mmol/L; [LDL-C] = 1.8[1.3–2.3] mmol/L). Further, a statistically significant association emerged for HDL-C and LDL-C levels on the day of the first diagnosis of ventilator-associated pneumonia (VAP) and mortality on day 28 paving the way for a deeper probing of lipoproteins and COVID-19. The kinetics of lipoproteins assayed in this work is depicted in figure 5 below (© 2020 Tanaka et al: *This is an open access article distributed under the terms of the Creative Commons Attribution License, which permits unrestricted use, distribution, and reproduction in any medium, provided the original author and source are credited*):

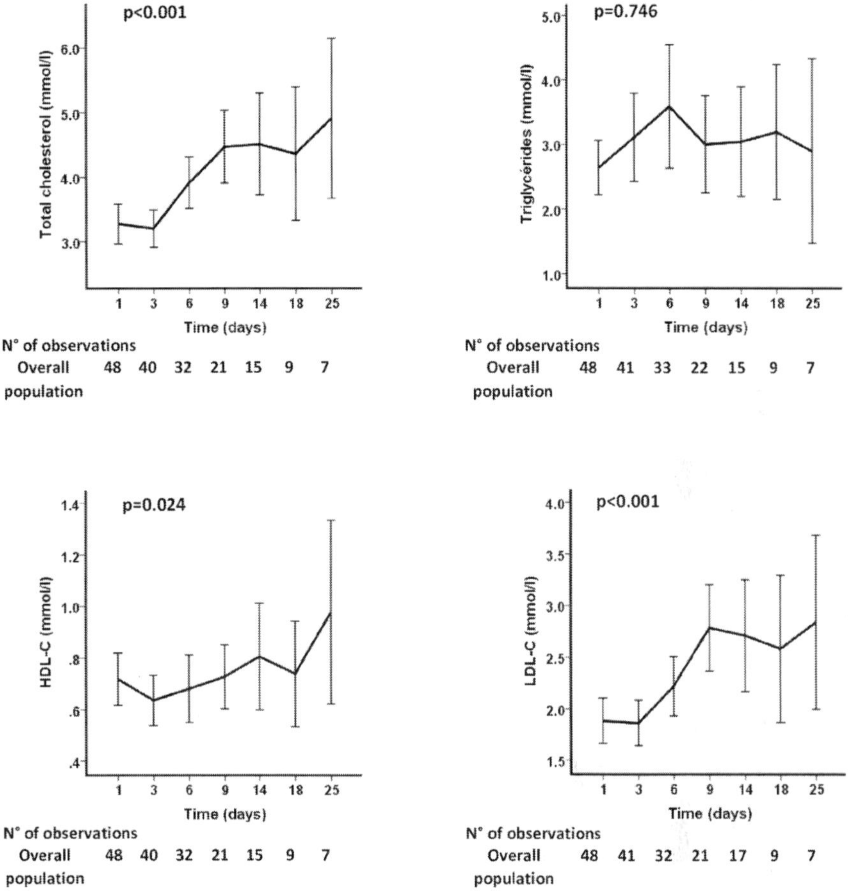

Another report by Masana and team scrutinized serum lipid levels in COVID-19 patients over the course of the infection. The work entailed an observational cross-sectional design with 2159 patients in the STACOV cohort (Cohort registration code NCT04407273-STACOV). The patient details are depicted in figure 6 below (Masana, L., Correig, E., Ibarretxe, D. et al, 2021: *Open Access: This article is licensed under a Creative Commons Attribution 4.0 International License, which permits use, sharing, adaptation, distribution and reproduction in any medium or format, as long as you give appropriate credit to the original author(s) and the source, provide a link to the Creative Commons licence, and indicate if changes were made. The images or other third party material in this article are included in the article's Creative Commonslicence, unless indicated otherwise in a credit line to the material. If material is not included in the article's Creative Commons licence and your intended use is not permitted by statutory regulation or exceeds the permitted use, you will need to obtain permission*

directly from the copyright holder. To view a copy of this licence, visit http://creativecommons.org/licenses/by/4.0/):

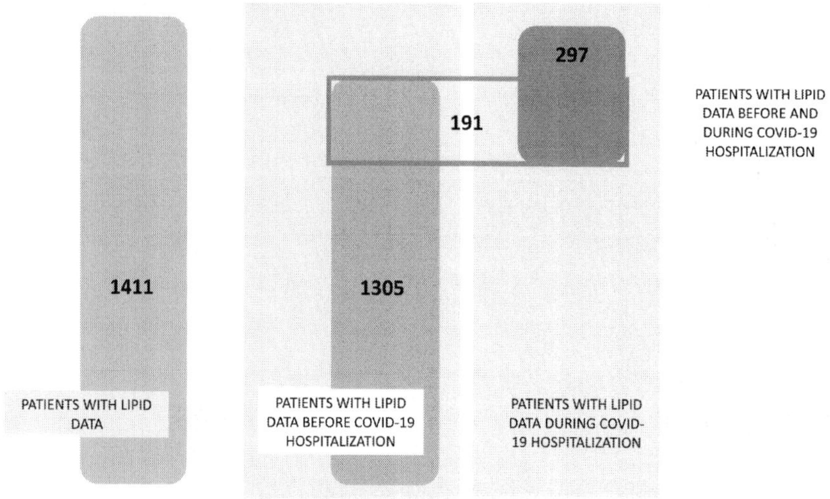

While both mild and severe cases were demonstrative of LDL-C levels on similar lines, HDL-C levels were lower and TG levels evidently higher in the severely infected patients. The levels of total cholesterol (TC), LDL-C, non-HDL-C and HDL-C were more in patients prior to infection and during hospitalization.

The lipid profile in COVID-19 infection is depicted below by the researchers in table 4 (Masana, L., Correig, E., Ibarretxe, D. et al, 2021: *Open Access: This article is licensed under a Creative Commons Attribution 4.0 International License, which permits use, sharing, adaptation, distribution and reproduction in any medium or format, as long as you give appropriate credit to the original author(s) and the source, provide a link to the Creative Commons licence, and indicate if changes were made. The images or other third party material in this article are included in the article's Creative Commonslicence, unless indicated otherwise in a credit line to the material. If material is not included in the article's Creative Commons licence and your intended use is not permitted by statutory regulation or exceeds the permitted use, you will need to obtain permission directly from the copyright holder. To view a copy of this licence, visit http://creativecommons.org/licenses/by/4.0/*):

Lipid profile available before COVID-19 hospitalization (median [IQR] in mmol/L)*				
	All **N = 1305**	**Mild** **N = 794**	**Severe** **N = 511**	**P-value**
Total cholesterol	4.84 [4.14; 5.54]	4,84 [4.17; 5.59]	4.82 [4.07; 5.46]	0.314
HDL cholesterol	1.3[1.09; 1.53]	1.32 [1.09; 1.58]	1.25[1.06; 1.48]	0.007
LDL cholesterol	2.85[2.25; 3.5]	2.87[2.25; 3.5]	2.85 [2.23; 3.42]	0.331
No-HDL cholesterol	3.5 [2.82; 4.2]	3.5 [2.8; 4.25]	3.52 [2.85; 4.1]	0.845
Triglycerides	1.37 [0.99; 1.87]	1.31 [0.95; 1.78]	1.44 [1.06; 1.99]	< 0.001
Lipid profile available during COVID-19 hospitalization				
	All **N = 297**	**Mild** **N = 149**	**Severe** **N = 148**	**P-value**
Total cholesterol	3.65 [3.03; 4.35]	3.63 [3.13; 163]	3.68 [2.97; 4.79]	0.528
HDL cholesterol	0.8 [0.64; 1.04]	0.88 [0.72; 1.08]	0.73 [0.59; 0.98]	< 0.001
LDL cholesterol	1.90 [1.39; 2.53]	1.86 [1.45; 2.40]	1.95 [1.32; 2.66]	0.757
No-HDL cholesterol	2.82 [2.15; 3.41]	2.79 [2.12; 3.26]	2.97 [2.19; 3.78]	0.033
Triglycerides	1.76 [1.22; 2.47]	1.61 [1.13; 2.18]	1.94 [1.39; 2.88]	< 0.001

During the hospital stay, a 16% drop in median HDL and 20% rise in TG was documented in severe infections. A poor prognosis emerged for the low HDL-C and high TG profile paving the way for these lipids as markers of infection. Figure 7 below illustrates ***TOP***: lipid profile changes in COVID-19 hospitalization (Masana, L., Correig, E., Ibarretxe, D. et al, 2021: *Open Access: This article is licensed under a Creative Commons Attribution 4.0 International License, which permits use, sharing, adaptation, distribution and reproduction in any medium or format, as long as you give appropriate credit to the original author(s) and the source, provide a link to the Creative Commons licence, and indicate if changes were made. The images or other third party material in this article are included in the article's Creative Commonslicence, unless indicated otherwise in a credit line to the material. If material is not included in the article's Creative Commons licence and your intended use is not permitted by statutory regulation or exceeds the permitted use, you will need to obtain permission directly from the copyright holder. To view a copy of this licence, visit http://creativecommons.org/licenses/by/4.0/*):

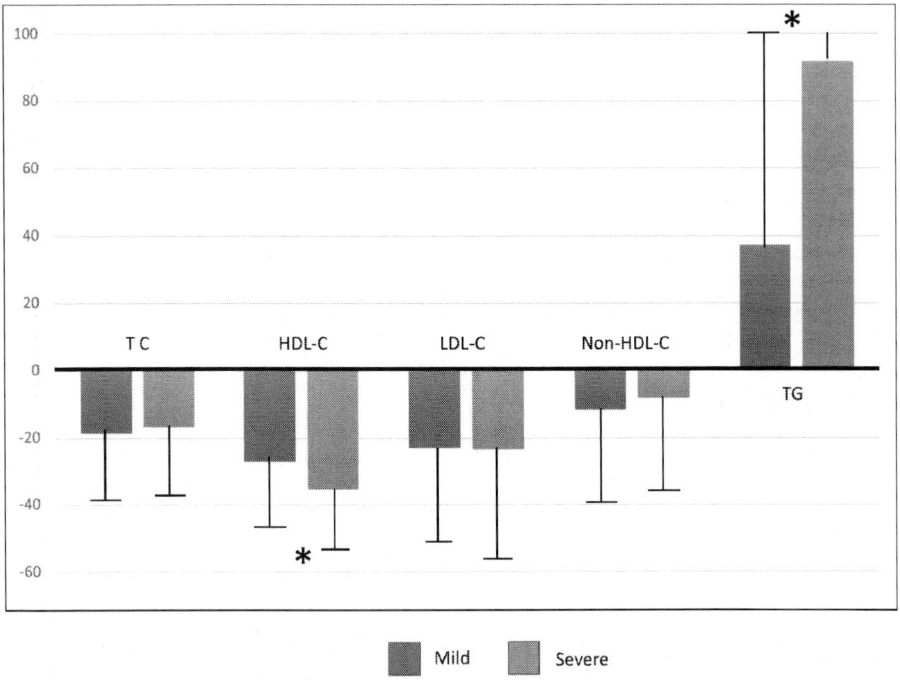

BOTTOM: Disease severity and TG/HDL profiles (Masana, L., Correig, E., Ibarretxe, D. et al, 2021: *Open Access: This article is licensed under a Creative Commons Attribution 4.0 International License, which permits use, sharing, adaptation, distribution and reproduction in any medium or format, as long as you give appropriate credit to the original author(s) and the source, provide a link to the Creative Commons licence, and indicate if changes were made. The images or other third party material in this article are included in the article's Creative Commonslicence, unless indicated otherwise in a credit line to the material. If material is not included in the article's Creative Commons licence and your intended use is not permitted by statutory regulation or exceeds the permitted use, you will need to obtain permission directly from the copyright holder. To view a copy of this licence, visit http://creativecommons.org/licenses/by/4.0/*):

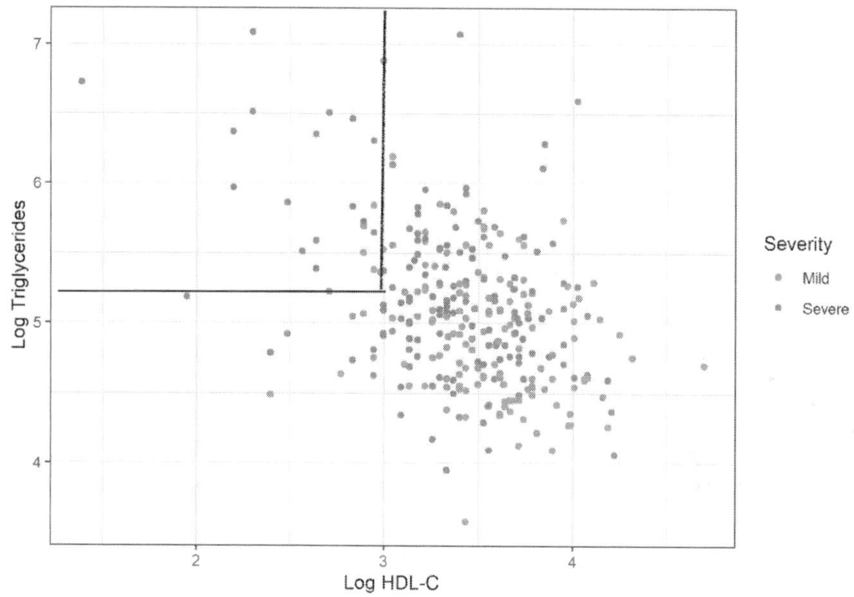

The analyses of HDL protein composition in COVID-19 patients employing mass spectrometry-based proteomics. Following the isolation of HDLs via ultracentrifugation of the plasma of severe COVID-19 patients on the day of their admission to the intensive care unit (ICU) (D1) and at D3 and D7 for survivors, the fractions were assayed by shotgun proteomics. Patients upon admission (D1) documented higher triglycerides and lowered total cholesterol, HDL-C and LDL-C levels vs. that of controls as documented in table 5 below (Begue, F., Tanaka, S., Mouktadi, Z. et al, 2021:):

TC, total cholesterol; TG, triglyceride; HDL-C, High-density lipoprotein cholesterol; LDL-C, Low-density lipoprotein cholesterol compared employing the Mann–Whitney test in controls (n = 16) vs. COVID-19 at admission (n = 8) patients.

Lipid profile	**Admission (D1)**	**Day 3**	**Day 7**	**Controls**	***P-values* (D1 vs Controls)**
TC, mmol/L, median [IQR]	2.65 [2.46–3.88]	2.68 [2.30–3.24]	4.09 [3.21–4.89]	5.25 [4.83–5.73]	0.0016
TG, mmol/L, median [IQR]	2.53 [1.73–2.99]	2.73 [2.0–3.18]	2.11 [2.02–3.60]	1.20 [0.90–1.83]	0.0152
HDL-C, mmol/L, median [IQR]	0.77 [0.49–0.83]	0.58 [0.44–0.86]	0.84 [0.64–0.91]	1.46 [1.34–1.69]	< 0.0001
LDL-C, mmol/L, median [IQR]	1.43 [1.28–2.20]	1.55 [1.13–2.0]	0.96 [0.90–2.19]	3.20 [2.15–3.65]	0.0081
Apo-AI, mg/dL, median [IQR]	48.3 [18.8–53.2]	45.1 [18.6–56.9]	56.9 [48.2–61.6]	92.8 [84.9–103.9]	< 0.0001

HDLs from COVID-19 patients demonstrated diminished apolipoproteins like ApoA-II, ApoC-I, II, III, IV, ApoA-IV, ApoC-I, ApoJ, Apo(a), ApoE, ApoM, ApoD, ApoB100 and ApoF with beta2-glycoprotein 1, CD44 antigen and serum paraoxonases 1 and 3 also documenting lowered levels. The level of paraoxonase 1 has been shown to be lowered in cases of sepsis.

The proteomic analyses of HDL particles from COVID-19 patients are documented below in figure 8 below: ***TOP***: Relative protein abundances (Begue, F., Tanaka, S., Mouktadi, Z. et al, 2021: *Open Access: This article is licensed under a Creative Commons Attribution 4.0 International License, which permits use, sharing, adaptation, distribution and reproduction in any medium or format, as long as you give appropriate credit to the original author(s) and the source, provide a link to the Creative Commons licence, and indicate if changes were made. The images or other third party material in this article are included in the article's Creative Commons licence, unless indicated otherwise in a credit line to the material. If material is not included in the article's Creative Commons licence and your intended use is not permitted by statutory regulation or exceeds the permitted use, you will need to obtain permission directly from the copyright holder. To view a copy of this licence, visit http://creativecommons.org/licenses/by/4.0/*):

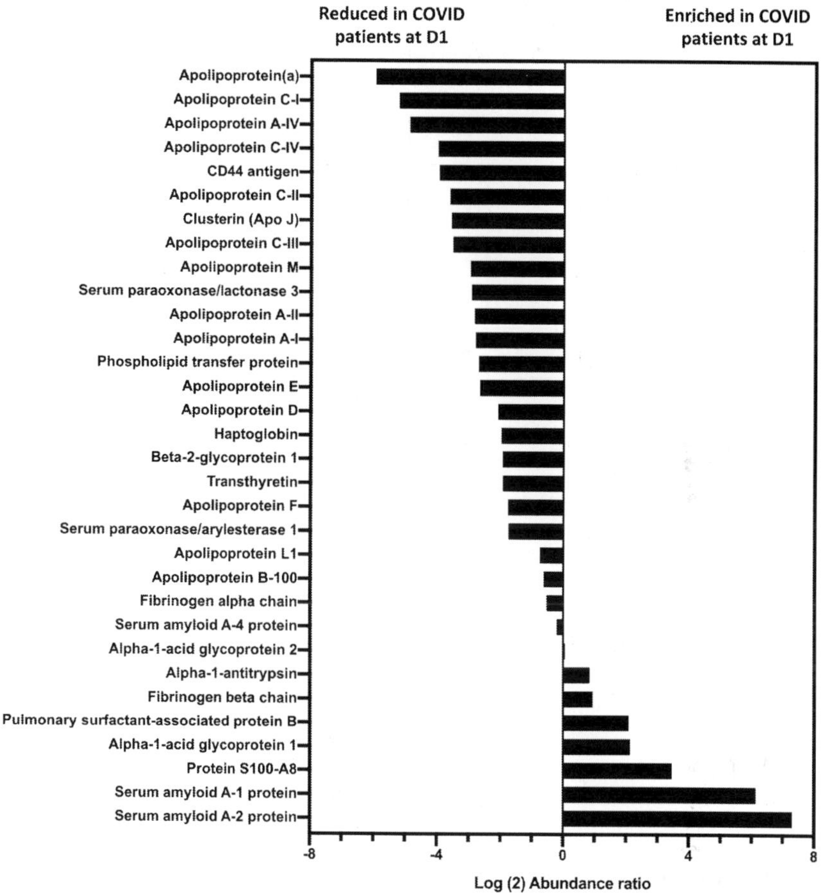

BOTTOM: Protein abundance (%) in controls vs. COVID-19 patient HDLs (Begue, F., Tanaka, S., Mouktadi, Z. et al, 2021: *Open Access: This article is licensed under a Creative Commons Attribution 4.0 International License, which permits use, sharing, adaptation, distribution and reproduction in any medium or format, as long as you give appropriate credit to the original author(s) and the source, provide a link to the Creative Commons licence, and indicate if changes were made. The images or other third party material in this article are included in the article's Creative Commons licence, unless indicated otherwise in a credit line to the material. If material is not included in the article's Creative Commons licence and your intended use is not permitted by statutory regulation or exceeds the permitted use, you will need to obtain permission directly from the copyright holder. To view a copy of this licence, visit http://creativecommons.org/licenses/by/4.0/*):

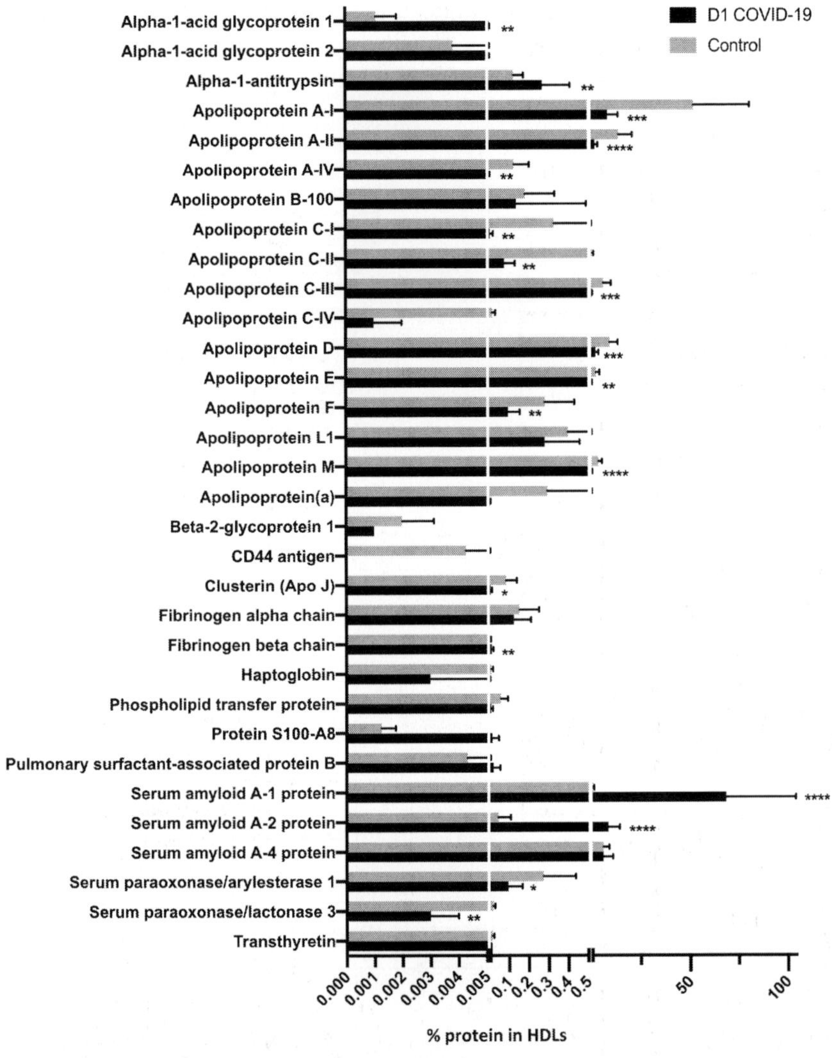

Further, as HDLs demonstrate protective properties on endothelial cells, the anti-inflammatory impact of HDLs from controls or patients following TNFα-based stimulation of HUVECs (human umbilical vein endothelial cells). The protective effect was lowered in COVID-19 patients with survivors however documenting better protection as opposed to non-survivor patients. As research has documented the protective effect of HDLs from apoptosis induced by TNFα, co-incubation with HDLs from control patients exerted a protective effect on HUVEC cell-cell interaction damage due to TNFα while COVID-19+ patient HDLs did not show this protective effect. Thus, apart from the abnormalities in HDL particles seen in COVID-19 patients, the loss of protection on endothelial cells is also an interesting observation.

Figure 9 below illustrates ***TOP***: Control and COVID-19 (C+) sample HDLs and TNFα-treated HUVECs: VE-Cadherin (red) and nuclear staining with DAPI (blue) while empty arrows show VE-Cadherin re-organization and plain arrows show apoptotic nuclei (Begue, F., Tanaka, S., Mouktadi, Z. et al, 2021: *Open Access: This article is licensed under a Creative Commons Attribution 4.0 International License, which permits use, sharing, adaptation, distribution and reproduction in any medium or format, as long as you give appropriate credit to the original author(s) and the source, provide a link to the Creative Commons licence, and indicate if changes were made. The images or other third party material in this article are included in the article's Creative Commons licence, unless indicated otherwise in a credit line to the material. If material is not included in the article's Creative Commons licence and your intended use is not permitted by statutory regulation or exceeds the permitted use, you will need to obtain permission directly from the copyright holder. To view a copy of this licence, visit http://creativecommons.org/licenses/by/4.0/*):

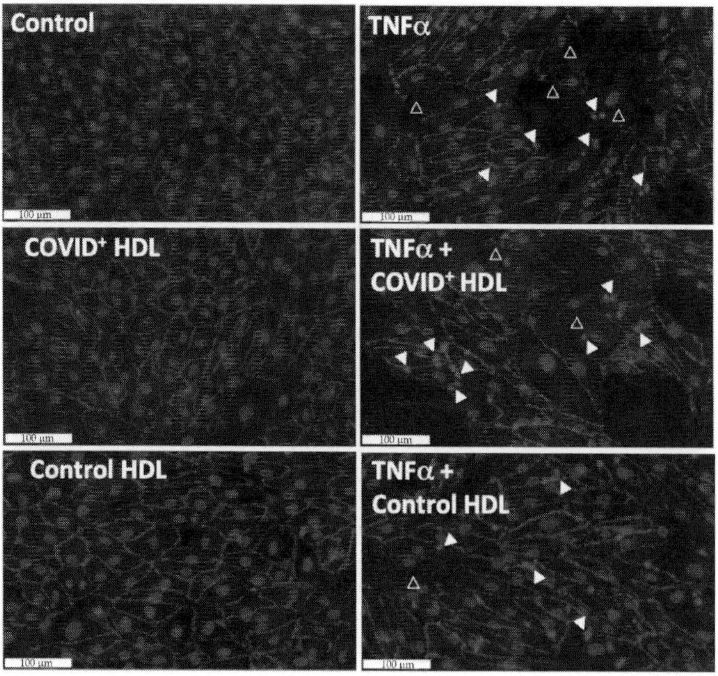

BOTTOM: Apoptosis scored by TUNEL and DAPI staining with nuclear condensation and fragmentation (DAPI in blue, A, C) and DNA fragmentation (green, B, C) while D is the means ± SD of apoptotic nuclei count (Begue, F., Tanaka, S., Mouktadi, Z. et al, 2021: *Open Access: This article is licensed under a Creative Commons Attribution 4.0 International License, which permits use, sharing, adaptation, distribution and reproduction in any medium or format, as long as you give appropriate credit to the original author(s) and the source, provide a link to the Creative Commons licence, and indicate if changes were made. The images or other third party material in this article are included in the article's Creative Commons licence, unless indicated otherwise in a credit line to the material. If material is not included in the article's Creative Commons licence and your intended use is not permitted by statutory regulation or exceeds the permitted use, you will need to obtain permission directly from the copyright holder. To view a copy of this licence, visit http://creativecommons.org/licenses/by/4.0/*):

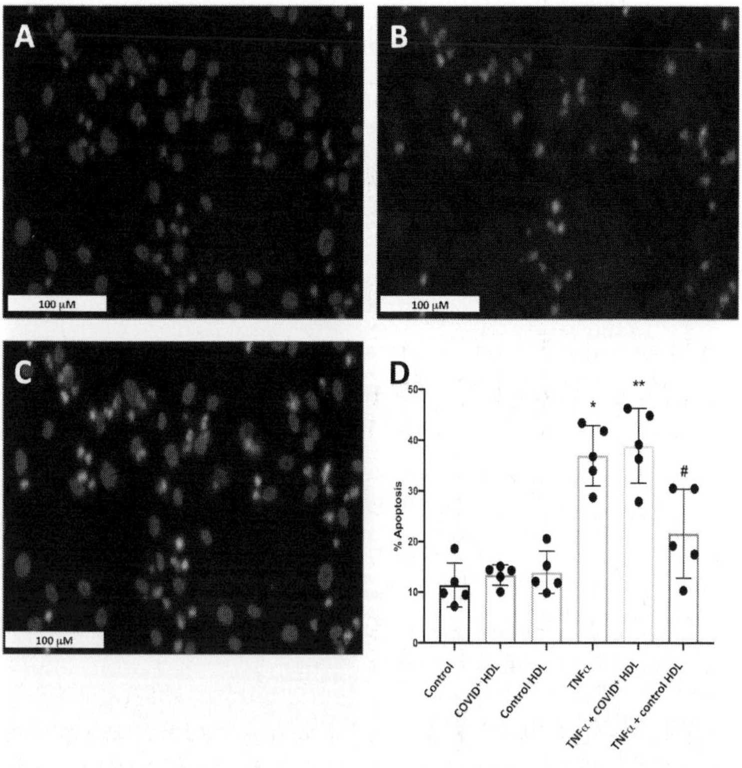

Scientists Vlasov and team (2021) published an RNA sequencing report (described as "for the first time to our knowledge") in *Cells* of patients admitted to the Intensive Care Unit (ICU) of the Pavlov First State Medical

University of St. Petersburg with COVID-19. The study entailed probing for differentially expressed genes (DEGs) from RNA isolated from peripheral blood mononuclear cells (PBMCs) as depicted in the scheme by the researchers in figure 10 below (Vlasov I, Panteleeva A, Usenko T, Nikolaev M, Izumchenko A, Gavrilova E, Shlyk I, Miroshnikova V, Shadrina M, Polushin Y, Pchelina S, Slonimsky P, 2021: © 2021 by the authors. Licensee MDPI, Basel, Switzerland. *This article is an open access article distributed under the terms and conditions of the Creative Commons Attribution (CC BY) license: https://creativecommons.org/licenses/by/4.0/*):

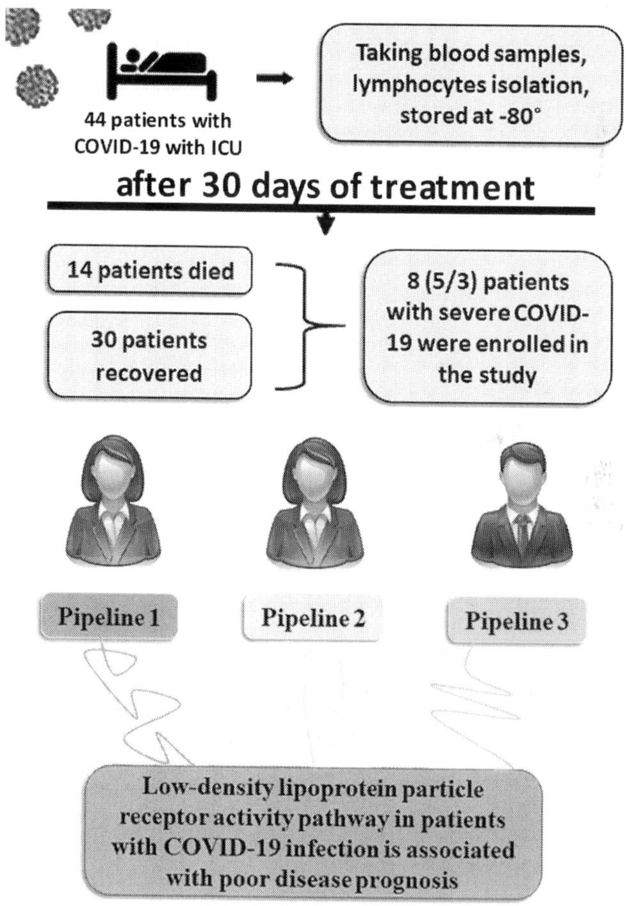

Subsequent Gene Ontology (GO) analysis revealed *three overlapping enriched pathways among the three pipelines: LDL particle receptor activity (GO:0005041), leukocyte differentiation (GO:0002521), cargo receptor activity (GO:0038024)*. The significant terms between the survivors and non-survivors were presented by the researchers as outlined in table 6 below

GO ID	GO term	padj	Associated Genes, %	Genes
Pipepline 1				
GO:0005041	low-density lipoprotein particle receptor activity	0,019	62.50	*ABCA2, ANXA2, CD36, LDLR, STAB1*
GO:0002521	leukocyte differentiation	0,015	14.79	*ANXA2, C1QC, CAMK4, CASP8, CBFA2T3, CD3D, CD74, CHD7, GAS6, HLA-DRB1, LCK, LGALS9, LILRB1, LILRB4, LY9, MS4A1, PDE1B, PPARG, RHOH, RRAS, TNFRSF11A*
GO:0038024	cargo receptor activity	0,0078	32.14	*ABCA2, ANXA2, CD36, DAB2, LDLR, LRP1, MRC1, SCARB2, STAB1*
Pipepline 2				
GO:0005041	low-density lipoprotein particle receptor activity	0,00030	75.00	*ABCA2, ANXA2, CD36, LDLR, LRP6, STAB1*
GO:0002521	leukocyte differentiation	0,0039	14.79	*AGER, ANXA2, C1QC, CAMK4, CASP8, CBFA2T3, CD3D, CHD7, CSF1, GAS6, LCK, LGALS9, LILRB1, LILRB4, LY9, MS4A1, PDE1B, PPARG, RHOH, RRAS, TNFRSF11A*
GO:0038024	cargo receptor activity	0,00036	35.71	*ABCA2, ANXA2, CD36, DAB2, LDLR, LRP1, LRP6, MRC1, SCARB2, STAB1*
(GO:0030228)	lipoprotein particle receptor activity	0,00439	54,55	*ABCA2, ANXA2, CD36, LDLR, LRP6, STAB1*
(GO:0045785)	positive regulation of cell adhesion	0,03722	11,71	*AFDN, AGER, CD1D, CD36, CD3E, CD6, DOCK1, DPP4, DUSP28, FLOT1, FN1, GCNT2, ITGAV, LCK, LGALS9, LILRB1, LILRB4, NCK2, NRG1, RASAL3, RHOH, RIN2, RNASE1, RRAS, S100A10, TEK*
(GO:0045321)	leukocyte activation	0,0281	10,53	*AGER, AHR, CAMK4, CD1D, CD22, CD33, CD3D, CD3E, CD6, CD84, CHD7, CXCL8, DPP4, EHHADH, GAS6,*

				GPER1, HAVCR2, IL10, LCK, LGALS3, LGALS9, LILRB1, LILRB4, LY9, MS4A1, NCK2, PRDX1, RASAL3, RHOH, SIT1, TREML2, VAMP8
Pipepline 3				
GO:0005041	low-density lipoprotein particle receptor activity	0,00015	46.15	*ANXA2, CD36, ITGAV, LRP6, PPARG, STAB1*
GO:0002521	leukocyte differentiation	0,0014	12.10	*ANXA2, C1QC, CBFA2T3, GAS6, IL10, LGALS1, LGALS9, LILRB1, LILRB4, LY9, PDE1B, PPARG, RHOH, RRAS, TNFRSF11A*
GO:0038024	cargo receptor activity	0,00002 2	31.03	*ANXA2, CD36, DAB2, ITGAV, LRP1, LRP6, PPARG, SCARB2, STAB1*
GO:0097048	dendritic cell apoptotic process	0,01071 21	75,00	*GAS6, LGALS9, LILRB1*
GO:0002573	myeloid leukocyte differentiation	0,03359 06	13.24	*ANXA2, C1QC, CBFA2T3, LGALS9, LILRB1, LILRB4, PDE1B, PPARG, TNFRSF11A*
GO:1903131	mononuclear cell differentiation	0,02380 98	12,66	*GAS6, IL10, LGALS1, LGALS9, LILRB1, LILRB4, LY9, PDE1B, PPARG, RHOH*
GO:2000668	regulation of dendritic cell apoptotic process	0,01071 21	75,00	*GAS6, LGALS9, LILRB1*
GO:0097028	dendritic cell differentiation	0,03720 9	33,33	*GAS6, LGALS1, LGALS9, LILRB1*
GO:0035590	purinergic nucleotide receptor signaling pathway	0,03720 9	33.33	*ADORA2A, P2RX4, P2RX7, P2RY1*

The identification of the LDL particle receptor pathway activity with up to 75% degree of enrichment in nonsurvivors becomes pertinent for our chapter. GO analysis for 361 DEGs overlapping the three pipelines unveiled that survivors documented downregulation of *STAB1*, *PPARG*, *CD36*, *ITGAV*, and *ANXA2* vs. the non-survivors. These genes are associated with cholesterol homeostasis: *CD36* and *STAB1* are scavenger receptors; *PPARG* regulates lipid metabolism; Annexin A2 (AnxA2) influences cholesterol metabolism among other functions. Further, the identification of the leukocyte pathway is suggestive of aberrant leukocyte differentiation in nonsurvivors. The GO network analysis of the GO terms (top enriched) in DEGs for survivors vs. nonsurvivors is depicted in figure 11 below (Vlasov I, Panteleeva A, Usenko T, Nikolaev M, Izumchenko A, Gavrilova E, Shlyk I, Miroshnikova V, Shadrina M, Polushin Y, Pchelina S, Slonimsky P, 2021:):

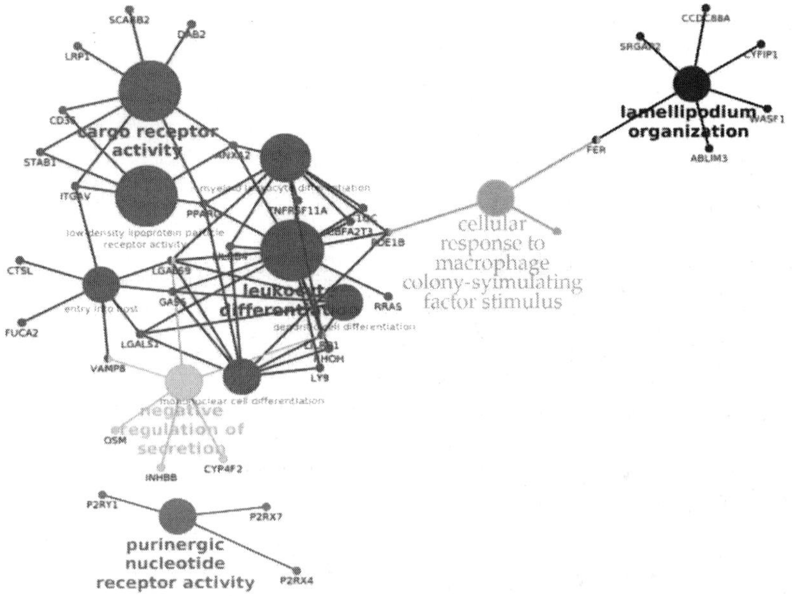

3.4 STUDIES EMPLOYING LIPIDOMICS

The lipidomic profile in the blood serum of COVID-19 hospitalized patients (n=53:Regione-Campania-cohort) was probed by targeted mass spectrometry (MS)-based along with the measurement few proinflammatory cytokines by Caterino and team (2021). The lipids scrutinized by the researchers were “diacylglycerols (DG) (n = 44 molecules), triacylglycerols (TG) (n = 242 molecules), glycerophospholipids (LPC, PC aa, PC ae) (n = 90 molecules), cholesterol esters (CE) (n = 22), ceramides (Cer) (n = 70) and sphingolipids (SM) (n = 15 molecules)”. The perturbations in the lipid profile in the patients are documented below in figure 12 below: ***TOP***: Mild, moderate and severe patient lipid profile A) Supervised partial least squares-discriminant analysis (PLS-DA) score plot and B) discriminant lipidic features with the lipid abundance displayed as 0 mild, 1 moderate and 2 severe patients C) Heatmaps of the same (Caterino, M., Gelzo, M., Sol, S. et al, 2021:):

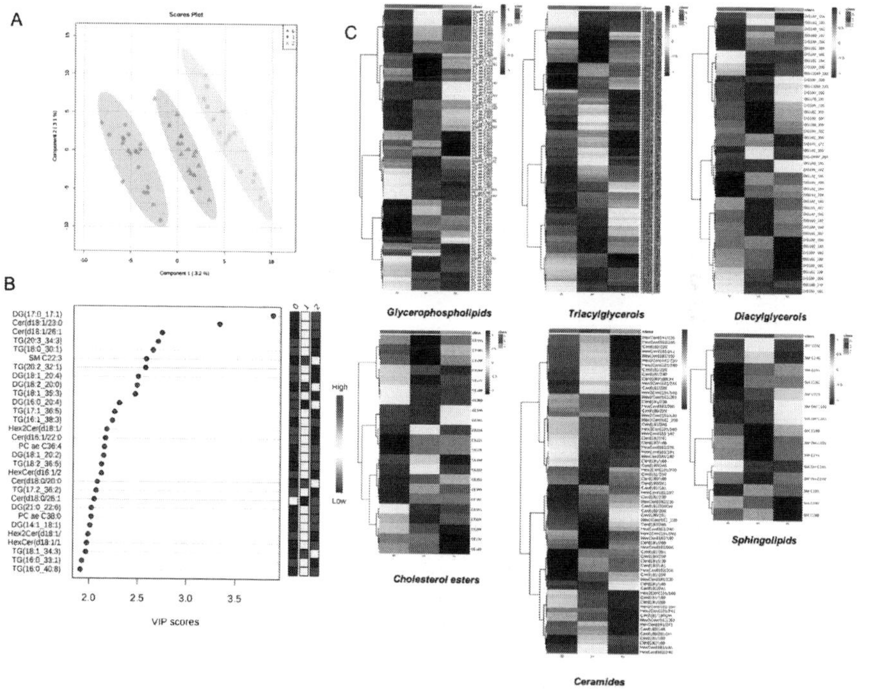

BOTTOM: Volcano plots of 483 lipids in (A) severe patient vs. mild and (B) severe patient vs. moderate. The fold change of significant differentially abundant lipids ($p < 0.05$) in (C) severe versus mild and (D) severe condition versus moderate patients (Caterino, M., Gelzo, M., Sol, S. et al, 2021:

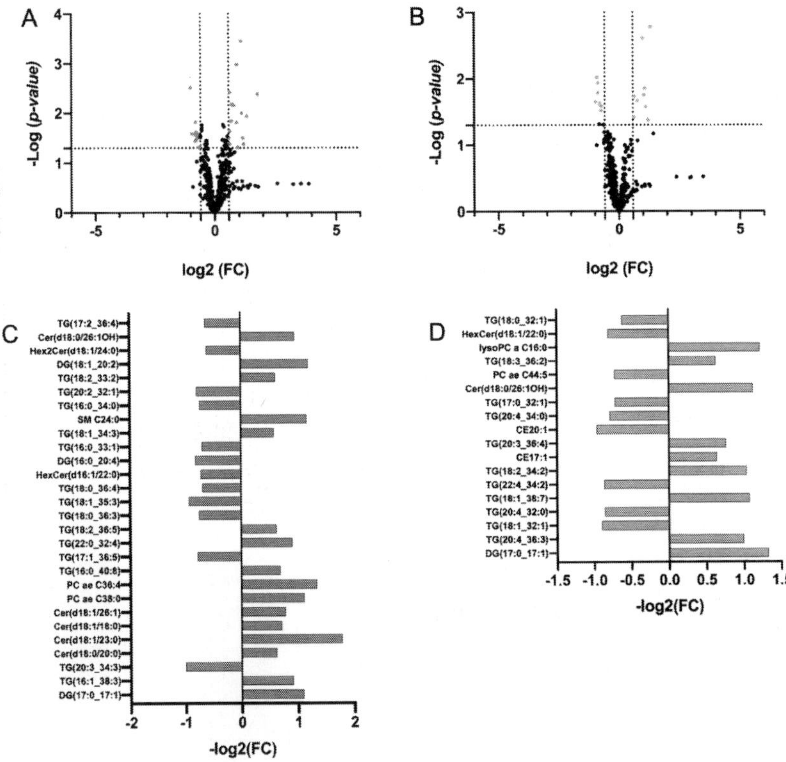

Meoni and team (2021) reported a "lipidomic signature" in patients (n=30) who were admitted at the Infective and Tropical Diseases and at the Intensive Care Unit of the Careggi University Hospital, Florence (IT) between March 10 and March 30 2020 employing untargeted nuclear magnetic resonance (NMR) spectroscopy of plasma-EDTA vs. relevant controls. Analyses unveiled an elevated VLDL particle profile and lowered Apo A1, Apo A2, cholesterol and free-cholesterol HDL and LDL subfractions in COVID-19 patients: conspicuous impacts were demonstrated in HDL-

3, HDL-4, LDL-4, LDL-5. Figure 13 below illustrates the lipidomic and metabolomic profiles in patients vs. controls: ***TOP***: Random Forest (RF) algorithm based categorization of COVID-19 patients (red dots) and controls (green dots) with A) metabolites and B) the lipoprotein-related parameters. (C) Confusion matrices. Log2 Fold Change (FC) of (D) quantified metabolites (E) lipoprotein-related parameters (© 2021 Meoni et al. *This is an open access article distributed under the terms of the Creative Commons Attribution License, which permits unrestricted use, distribution, and reproduction in any medium, provided the original author and source are credited*):

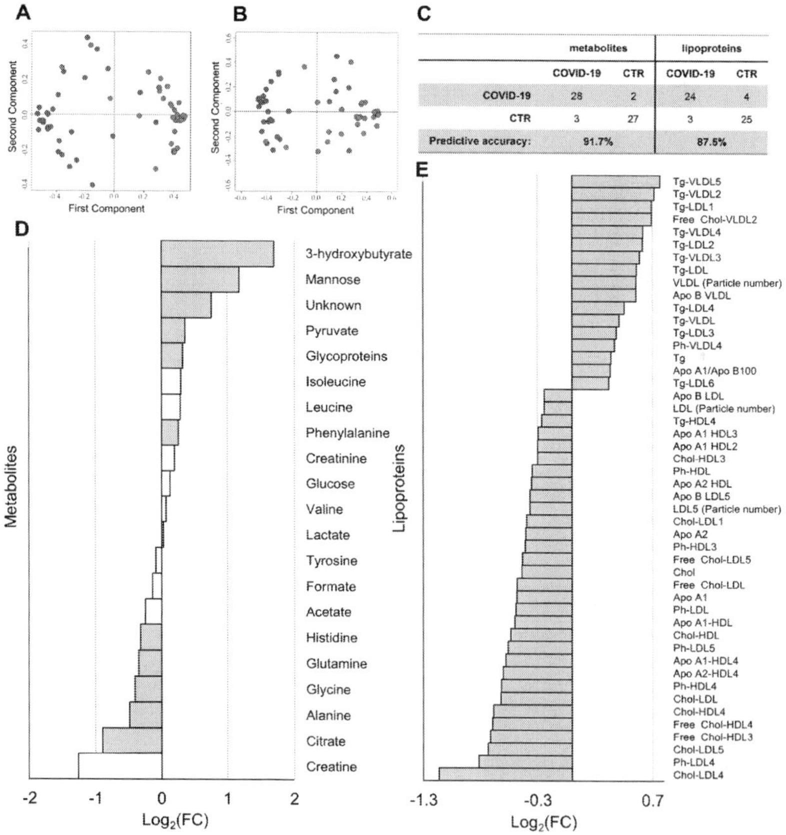

BOTTOM: Heatmap of assayed parameters: red, positive correlations; blue, negative correlations (© 2021 Meoni et al. *This is an open access article distributed under the terms of the Creative Commons Attribution License, which permits unrestricted use, distribution, and reproduction in any medium, provided the original author and source are credited*):

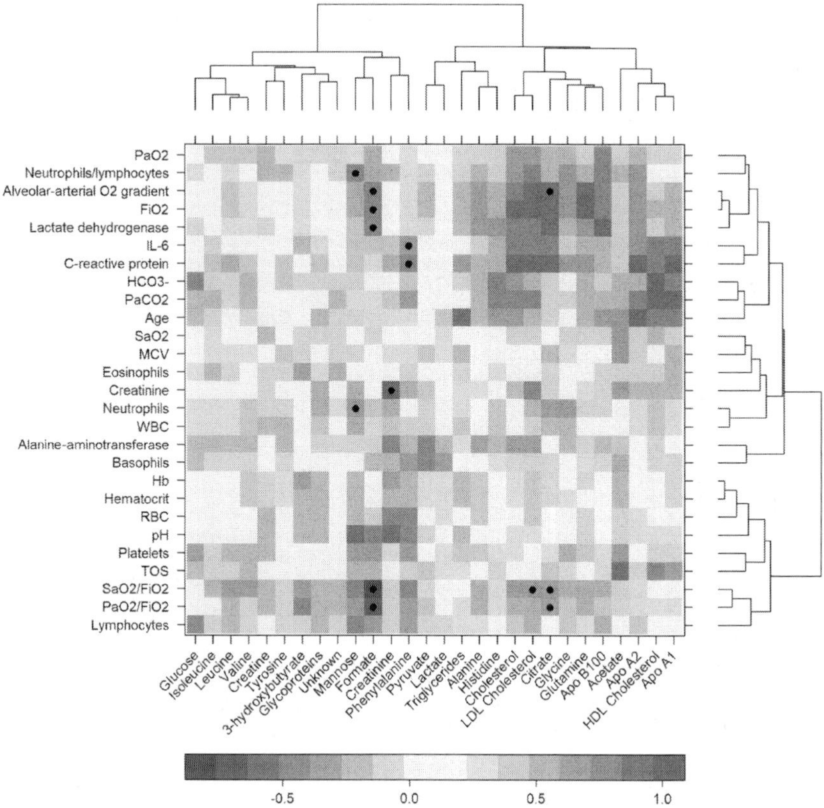

The lipoprotein and metabolomics profile in COVID-19 patients hospitalized at the intensive care unit (ICU) of the University Hospital of Schleswig-Holstein (Campus Lübeck) employing NMR metabolomics was reported by Schmelter and team (2021) in *Frontiers in Molecular Sciences*. This was compared with the profile generated for healthy controls (HCs) and a cohort of cardiogenic shock (CS) patients treated in the same ICU who tested negative for SARS-CoV-2. COVID-19 patients documented an aberrant energy status with elevated glucose and formic acid with lowered lactic acid and the lactic acid/pyruvic acid ratio vs. controls in terms of metabolites. Severe dyslipidemia was unveiled in the patients as evidenced by a conspicuous rise in the ratio of apolipoprotein B100 to apolipoprotein A1 (ApoB100/ApoA1: cardiovascular disease risk factor) and triglycerides, phospholipids, esterified, and free cholesterol bound to lipoproteins of smaller sizes (VLDL-3, 4, and 5) with lowered profiles of LDL and HDL.

Figure 14 below documents the perturbed metabolites (A) and lipoprotein classes (B-I) in COVID-19 patients. The middle line shows the reference average; circles and diamonds on horizontal axes document changes scaled by standard deviations. HCs are vertical lines; blue circles are CS and red circles are for COVID-19 patients with bold fonts highlighting the conspicuous alterations ():

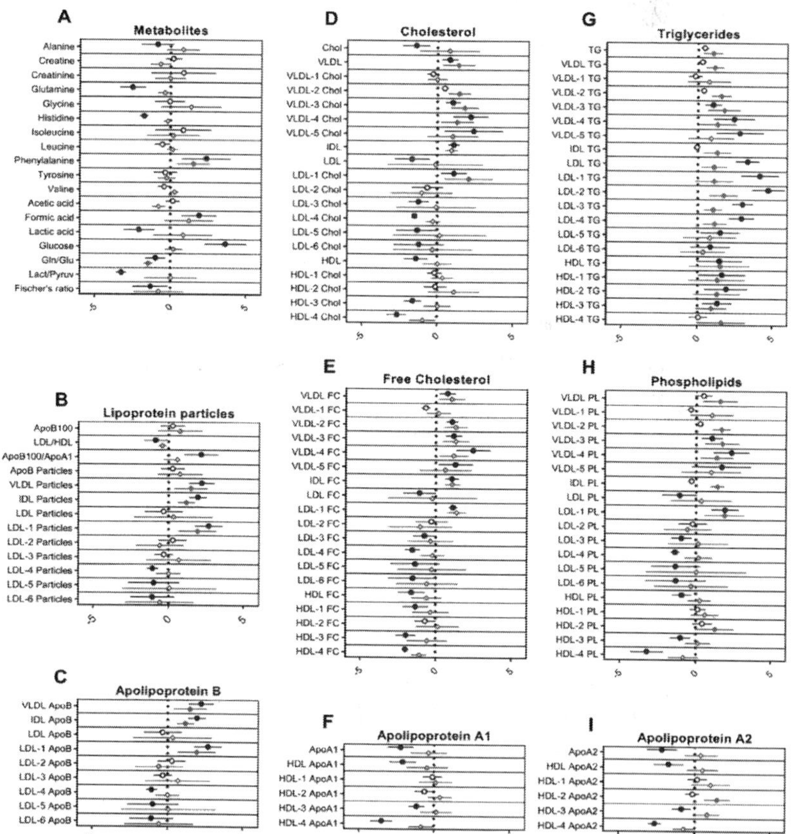

Scoring of alterations in mild/asymptomatic conditions entailed probing of antibody-positive individuals: anti-S1-IgG+ vs. anti-S1-IgG-(both PCR-negative for asymptomatic SARS-CoV-2 infection). A negative correlation emerged for small-sized HDL-4, small-sized LDL-6 particles and LDL-6 bound triglycerides, cholesterol, esterified cholesterol, and phospholipids with antibody titers. Figure 15 below illustrates the (A) Heatmap of metabolite and lipoprotein levels in anti-S1-IgG+(B–P) Spearman correlation plots for most significant individual metabolites or lipoproteins against anti-S1-IgG titer (© 2021 Schmelter, Föh, Mallagaray, Rahmöller, Ehlers, Lehrian, von Kopylow, Künsting, Lixenfeld, Martin, Ragab, Meyer-Saraei, Kreutzmann, Eitel, Taube, Käding, Jantzen, Graf, Sina and Günther. *This is an open-access article distributed under the terms of the Creative Commons Attribution License (CC BY). The use, distribution or reproduction in other forums is permitted, provided the original author(s) and the copyright owner(s) are credited and that the original publication in this journal is cited, in accordance with accepted academic practice. No use, distribution or reproduction is permitted which does not comply with these terms*):

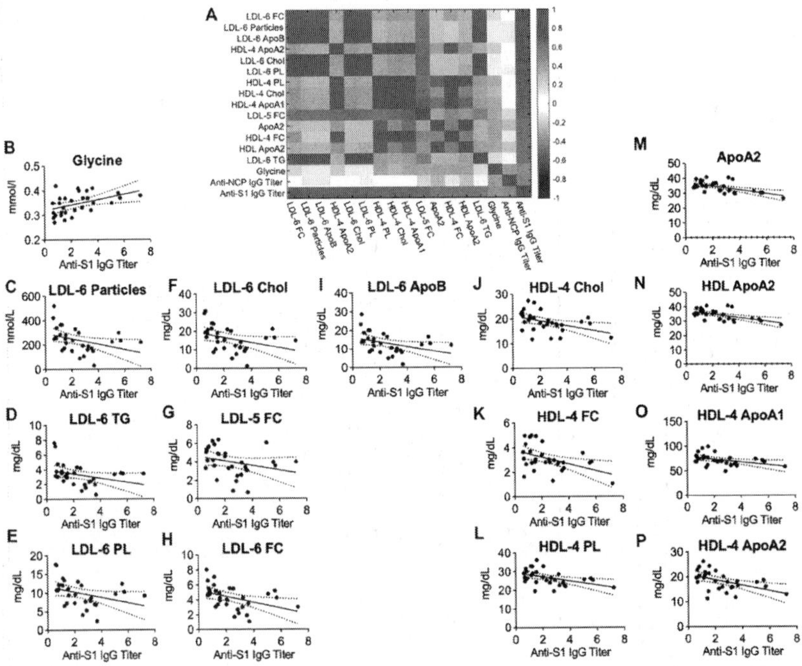

3.5 STUDIES PROBING SIGNATURES

3.5.1 Multiomics Signature inclusive of Cholesterol

The quest for suitable interventions to identify molecular patterns led to Yuming Li and team (2021) to probe protein, amino acid, and lipid (multiomics) in plasma and urine samples of covid-19 patients (severe and non-severe) vs. healthy donors as controls employing LC–MS and published the analyses in *signal Transduction and Targeted Therapy*. The training data set encompassed non-severe patients, severe patients and healthy volunteers (n=10 each) that were analyzed as depicted in figure 16 below: a) Quantification of proteins, lipids, and amino acids of COVID-19 patients in the training dataset time point details (Li et al, 2021: *Open Access: This article is licensed under a Creative Commons Attribution 4.0 International License, which permits use, sharing, adaptation, distribution and reproduction in any medium or format, as long as you give appropriate credit to the original author(s) and the source, provide a link to the Creative Commons license, and indicate if changes were made. The images or other third party material in this article are included in the article's Creative Commons license, unless indicated otherwise in a credit line to the material. If material is not included in the article's Creative Commons license and your intended use is not permitted by statutory regulation or exceeds the permitted use, you will need to obtain permission directly from the copyright holder. To view a copy of this license, visit http://creativecommons.org/licenses/by/4.0/*):

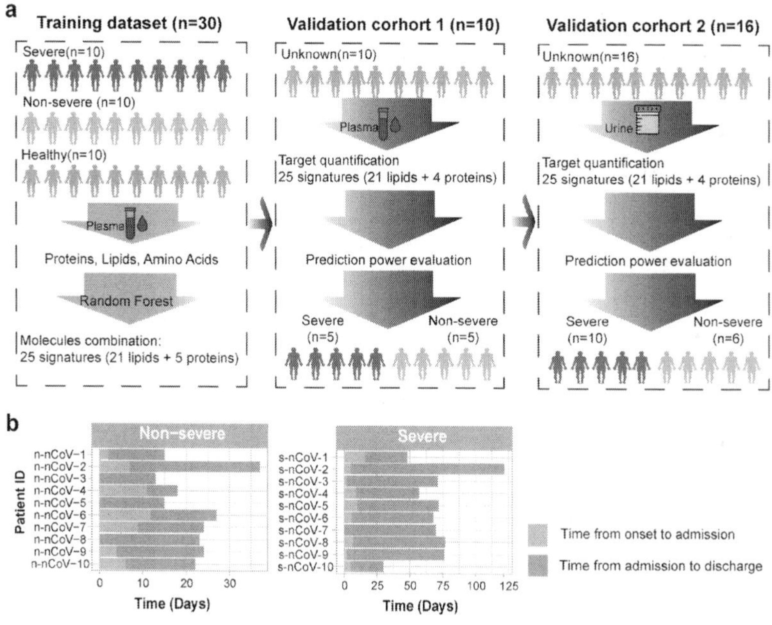

Of 1254 proteins scored, differentially expressed proteins among the datasets were associated with immune and infectious pathways with the severe and non-severe groups documenting 105 dysregulated proteins (34 upregulated and 71 downregulated) while the COVID-19 (non-severe and severe) and healthy groups documented 118 dysregulated proteins (86 upregulated and 32 downregulated). The non-severe vs. healthy group and the severe vs. healthy group comparison revealed 104 dysregulated proteins (76 upregulated and 28 downregulated) and 143 dysregulated proteins (80 upregulated and 63 downregulated), respectively.

Detailed scrutiny revealed that an enrichment of chemokine C–C motif ligand 18 (CCL18: involved in immune modulation), C-reactive protein (CRP: an acute inflammatory protein), and cholesteryl ester transfer protein (CETP: cholesteryl ester transport that was lowered in infected individuals as opposed to controls).

Figure 17 below documents ***TOP***: Proteomic analyses of the datasets a) Volcano plots of quantified proteins b) Heatmaps; FC-fold-change (Li et al, 2021: *Open Access: This article is licensed under a Creative Commons Attribution 4.0 International License, which permits use, sharing, adaptation, distribution and reproduction in any medium or format, as long as you give appropriate credit to the original author(s) and the source, provide a link to the Creative Commons license, and indicate if changes were made. The images or other third party material in this article are included in the article's Creative Commons license, unless indicated otherwise in a credit line to the material. If material is not included in the article's Creative Commons license and your intended use is not permitted by statutory regulation or exceeds the permitted use, you will need to obtain permission directly from the copyright holder. To view a copy of this license, visit http://creativecommons.org/licenses/by/4.0/*):

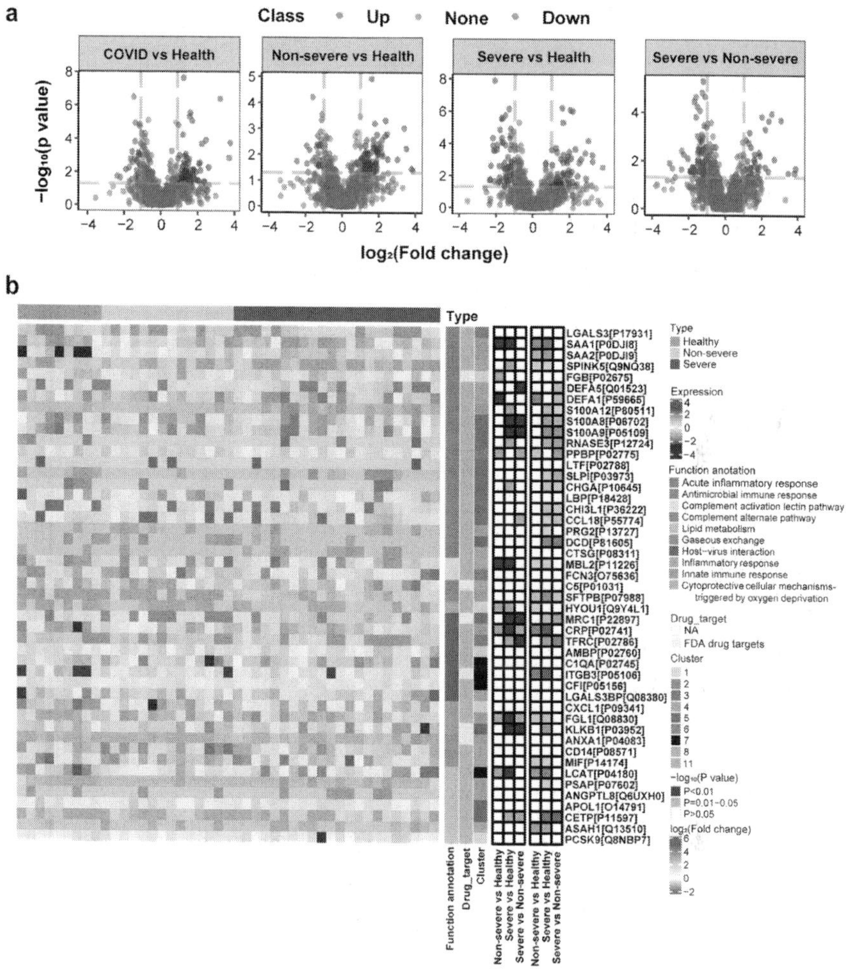

BOTTOM: Differential protein enrichment (A) between COVID-19 and Healthy groups. (B) between Non-severe and Healthy groups. (C) between Severe and Healthy groups. (D) Pathways enriched for differential proteins between Severe and Non-severe groups (Li et al, 2021:

permitted use, you will need to obtain permission directly from the copyright holder. To view a copy of this license, visit http://creativecommons.org/licenses/by/4.0/):

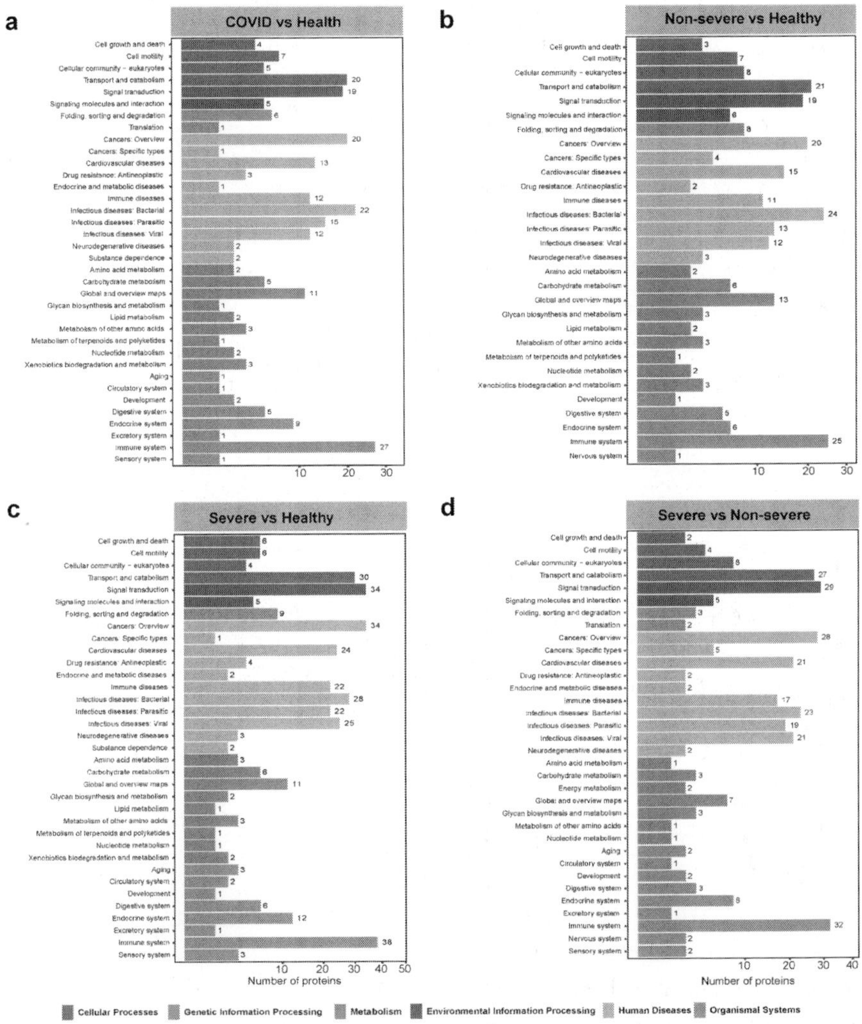

Plasma lipids documented alterations on the lines of inflammation with elevated phosphoserine (PS) and lowered phosphoglycerol (PG) in SARS-CoV-2 infection. Further, a panel of molecules was constructed with a cholesterol ester (ChE (18:1)) and other molecules inclusive of ceramides, glycosylceramides, phosphatidylethanolamines, etc (25 molecular signatures). The AUCreached 0.988 with 95% CI (0.75–1) in the ROC analyses for this signature in categorizing severe and non-severe patients. Further, these various factors assessed were developed as a model to account for non-severe and severe COVID-19vs. healthy controls as depicted in figure 18 below (A) SARS-CoV-2 and protein/lipid factors (B) Protein, lipid and amino acid alterationsforCOVID-19 (Li et al, 2021: *Open Access: This article is licensed under a Creative Commons Attribution 4.0 International License, which permits use, sharing, adaptation, distribution and reproduction in any medium or format, as long as you give appropriate credit to the original author(s) and the source, provide a link to the Creative Commons license, and indicate if changes were made. The images or other third party material in this article are included in the article's Creative Commons license, unless indicated otherwise in a credit line to the material. If material is not included in the article's Creative Commons license and your intended use is not permitted by statutory regulation or exceeds the permitted use, you will need to obtain permission directly from the copyright holder. To view a copy of this license, visit http://creativecommons.org/licenses/by/4.0/*):

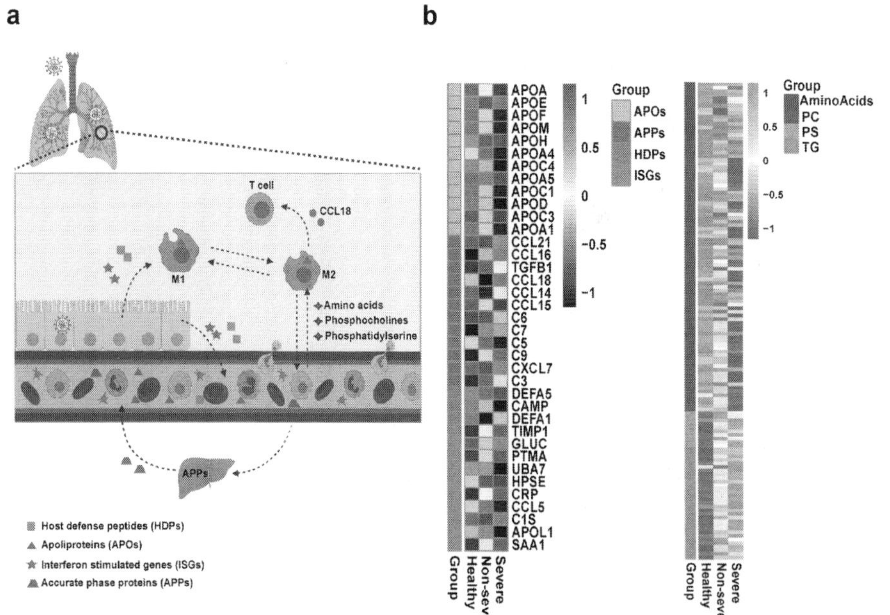

3.5.2 A Pilot NMR Investigation: The NIH Lipo-COVID Study

Researchers Ballout and team (2021) probed the alterations in lipoproteins and metabolites in consenting individuals (n=32)diagnosed with severe COVID-19 and were admitted to the intensive care units (ICU) of the Johns Hopkins University Hospital and/or the University of Maryland Medical

Center, Baltimore, MD, USA, entailing intubation and extracorporeal membrane oxygenation (ECMO). The study conducted at the Clinical Center of the National Institutes of Health (NIH) unveiled that 93.8% of patients documented a lowering of small HDL particles (S-HDL-P) with 90% of patients also showing a conspicuous lowering of HDL particles (HDL-P). 43.8% of the patients demonstrated increased TG-rich lipoprotein particles (TRL-P) levels.

Table 7 below documents the lipoprotein profile in COVID-19 patients as scored by NMR (Ballout RA, Kong H, Sampson M, Otvos JD, Cox AL, Agbor-Enoh S, Remaley AT, 2021: © 2021 by the authors.Licensee MDPI, Basel, Switzerland. *This article is an open access article distributed under the terms and conditions of the Creative Commons Attribution (CC BY) license: https://creativecommons.org/licenses/by/4.0/*):

Lipid Parameters	Number *(%)* of Subjects Outside Reference Range	Mean	Median	Interquartile Range (25th–75th Percentile)	Reference Range
HDL-P	29 (90.6) ↓	7 μmol/L	6.4 μmol/L	4.7–9.3 μmol/L	15.2–27.5 μmol/L
L-HDL-P	2 (6.3) ↓	1	1.3	0.88–1.8	0.1–6.9
M-HDL-P	12 (37.5) ↓	3	2.7	0.91–4.4	1.6–8.1
S-HDL-P	30 (93.8) ↓	3	1.6	0.18–4.6	8.2–20.6
TRL-P	14 (43.8) ↑	275 nmol/L	277 nmol/L	155–349 nmol/L	19–291 nmol/L
VL-TRL-P	1 (3.1) ↑	0	0.03	0.001–0.15	0–0.7
L-TRL-P	5 (15.6) ↑	8	3.64	0.063–8.06	0–11.9
M-TRL-P	12 (37.5) ↓	25	11	0–31	0.4–70.7
S-TRL-P	10 (31.3) ↑	79	69	25–116	0–103.6
VS-TRL-P	11 (34.4) ↑	164	154	65–239	0–184.2
LDL-P	4 (12.5) ↑	1890 nmol/L	1800 nmol/L	1397–2305 nmol/L	851–2585 nmol/L
L-LDL-P	0	63	7.1	0–90	0–674
M-LDL-P	0	61	0	0–35	0–1376
S-LDL-P	14 (43.8) ↓	549	108	0–811	71–1865
LpZ particles	25 (78.1) ↑	1217 nmol/L	1280 nmol/L	594–1907 nmol/L	0 nmol/L
LpX particles	23 (72) ↑	149 mg/dL	191 mg/dL	0–253 mg/dL	0 mg/dL

Notwithstanding lowered L-LDL-P and M-LDL-P (large and medium-sized LDL particles), the total LDL particle numbers (LDL-P) were in the range. Close to 80% of patients documented lipoprotein Z particles (LpZ) or lipoprotein B particles described as LDL particles enriched in TG and are not detected in healthy people but have been observed in alcoholic liver disease. Another abnormal lipoprotein particle called lipoprotein X particle (LpX) was reported in more than 70% of the patients; this particle is seen more in LCAT deficiency or cholestatic liver disease. Table 8 below documents these findings of altered lipoprotein levels as per the BMI: healthy weight (BMI < 25 kg/m2) and obese patients (BMI ≥ 30 kg/m2) (Ballout RA, Kong H, Sampson M, Otvos JD, Cox AL, Agbor-Enoh S, Remaley AT, 2021: © 2021 by the authors. Licensee MDPI, Basel, Switzerland. *This article is an open access article distributed under the terms and conditions of the Creative Commons Attribution (CC BY) license: https://creativecommons.org/licenses/by/4.0/*):

Body Mass Index (Kg/m²)	**Lipid Parameters**	**Mean**	**Median**	**Interquartile Range (25th–75th Percentile)**	**Reference Range**
<25 (N = 3)	**HDL-P**	13 µmol/L	15.5 µmol/L	10.4–16.4 µmol/L	15.2–27.5 µmol/L
	L-HDL-P	1	1.6	1.2–1.8	0.1–6.9
	M-HDL-P	6	3.7	2.6–8.7	1.6–8.1
	S-HDL-P	5	3	2.2–6.6	8.2–20.6
	TRL-P	180 nmol/L	157 nmol/L	107–242 nmol/L	19–291 nmol/L
	VL-TRL-P	0	0.1	0.1–0.4	0–0.7
	L-TRL-P	3	3.9	2–5.1	0–11.9
	M-TRL-P	18	21	10–27	0.4–70.7
	S-TRL-P	45	63	34–65	0–103.6
	VS-TRL-P	113	55	53–144	0–184.2
	LpZ particles	394 nmol/L	0 nmol/L	0–592 nmol/L	0 nmol/L
	LpX particles	0 mg/dL	0 mg/dL	0 mg/dL	0 mg/dL
≥30 (N = 21)	**HDL-P**	7 µmol/L	6.92 µmol/L	4.3–9.9 µmol/L	15.2–27.5 µmol/L
	L-HDL-P	2	1.4	0.99–1.9	0.1–6.9
	M-HDL-P	2	1.61	0.62–4.1	1.6–8.1
	S-HDL-P	3	2.73	0.06–6.5	8.2–20.6
	TRL-P	278 nmol/L	270.5 nmol/L	166–346 nmol/L	19–291 nmol/L

	VL-TRL-P	0	0.058	0.002–0.152	0–0.7
	L-TRL-P	10	3.43	0.26–10.4	0–11.9
	M-TRL-P	26	4.78	0–27.5	0.4–70.7
	S-TRL-P	80	76.3	25.4–116	0–103.6
	VS-TRL-P	162	151	67–228	0–184.2
	LpZ particles	1268 nmol/L	1378 nmol/L	615–1991 nmol/L	0 nmol/L
	LpX particles	131 mg/dL	172 mg/dL	2.89–247 mg/dL	0 mg/dL

These observations are interesting as S-HDL-P serves anti-oxidant and anti-inflammatory functions-the lowering of this lipoprotein in COVID-19 patients is noteworthy as the risk of cardiovascular disease (CVD) and CVD-associated mortality is more with patients with lowered S-HDL-P. Further, increased TRL-P has been reported in CVD and all-cause mortality while abnormal lipoprotein particles (LpZ and LpX) linked to liver damage also pave the way for further understanding the impact of COVID-19.

3.6 COVID-19 VACCINATION AND THE CHOLESTEROL PROFILE

While predominantly research has cast its light on antibody generation following vaccination, its impact on other aspects of physiology warrants probing. This emerges pertinent with globally large-scale vaccinations across nations necessitating the scrutiny of how newly developed vaccines impact our systems. SARS-CoV-2 infections were linked to a slew of respiratory maladies along with gastrointestinal manifestations and also complications inclusive of blood vessel issues, clotting disorders, kidney issues, electrolytes imbalances, to name a few. Given that the extent of mimicking of a COVID-19 infection by a vaccine warrants comprehensive elucidation, researchers Liu and team (2021) scored the impact of an inactivated SARS-CoV-2 vaccine (Vero Cell) as published in *Cell Discovery*. Apart from scoring the impact on immune cells employing large-scale single-cell mRNA sequencing (scRNA-seq), laboratory parameters were also scrutinized.

Figure 19 below illustrates the scheme employed to score the impact of the vaccination. **a** Illustrative description. **b**, **c** Inoculation of volunteers of cohort A (day 1 & day 14) and of cohort B (day 1 & day 28) with changes in antibodies documented in red (IgM) and black (IgG). **d** Alterations in neutralizing antibody titer post-vaccination and post-recovery in positive individuals (Liu et al, 2021: *Open Access: This article is licensed under*

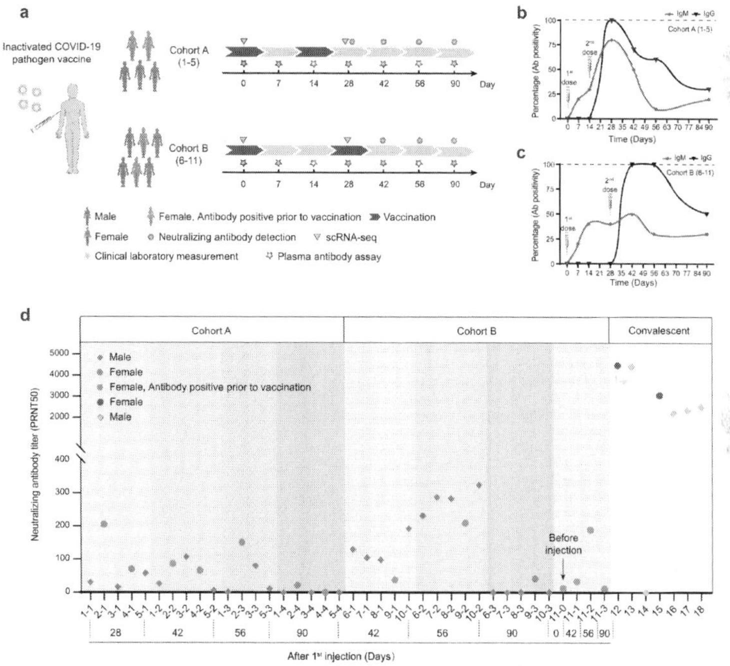

The immune responses were on the lines of an actual infection with elevated NF-κB-linked inflammatory signaling and monocytes while alterations in cholesterol and glucose profiles emerged as if these people had experienced an infection with SARS-CoV-2. In terms of cholesterol, boosted levels were documented on days 7 and 28 subsequent to the 1st inoculation.

Figure 20 below depicts the alterations in various factors: **a** A slew of tests were conducted inclusive of white blood cell count (**b**), HbA1c (**c**), potassium (**d**, left panel), sodium (**d**, right panel), APTT (**e**, left panel), PT (**e**, right panel), total cholesterol (**f**), total bile acid (**g**), eGFR (**h**, left panel), creatinine (**h**, right panel). The normal limits are depicted as horizontal dashed lines: red (upper) and blue (lower) (Liu et al, 2021: *Open Access: This article is licensed under a Creative Commons Attribution 4.0 International License, which permits use, sharing, adaptation, distribution and reproduction in any medium or format, as long as you give appropriate credit to the original author(s) and the source, provide a link to the Creative Commons license, and indicate if changes were made. The images or other third party material in this article are included in the article's Creative Commons license, unless indicated otherwise in a credit line to the material. If material is not included in the article's Creative Commons license and your intended use is not permitted by statutory regulation or exceeds the permitted use, you will need to obtain permission directly from the copyright holder. To view a copy of this license, visit http://creativecommons.org/licenses/by/4.0/*):

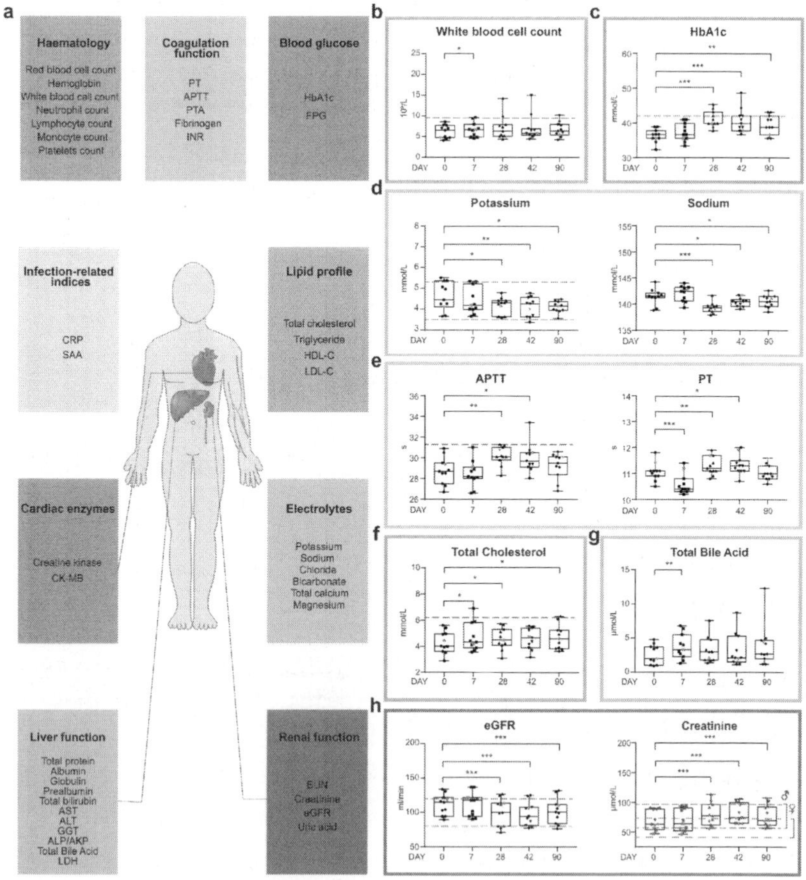

Analyses of differentially expressed genes (DEGs) unveiled an upregulation of genes on the lines of the above-mentioned parameters:

"cholesterol homeostasis", "coagulation", and "inflammatory response" (CXCL8, CD14, IL6, and TNFRSF1B), "TNFα signaling via NF-κB" (NFKB1, NFKB2, NFKBIE, TNFAIP3, and TNFSF9) and "hypoxia" (HIF1A) along with "TGFβ signaling", "IL2-STAT5 signaling" (IFNGR1, MAPKAPK2, and CASP3), and "IL6-JAK-STAT3 signaling"-related genes. The alterations in the immune cells in terms of DEGs post-vaccination are depicted in figure 21 below (Liu et al, 2021: *Open Access: This article is licensed under a Creative Commons Attribution 4.0 International License, which permits use, sharing, adaptation, distribution and reproduction in any medium or format, as long as you give appropriate credit to the original author(s) and the source, provide a link to the Creative Commons license, and indicate if changes were made. The images or other third party material in this article are included in the article's Creative Commons license, unless indicated otherwise in a credit line to the material. If material is not included in the article's Creative Commons license and your intended use is not permitted by statutory regulation or exceeds the permitted use, you will need to obtain permission directly from the copyright holder. To view a copy of this license, visit http://creativecommons.org/licenses/by/4.0/*):

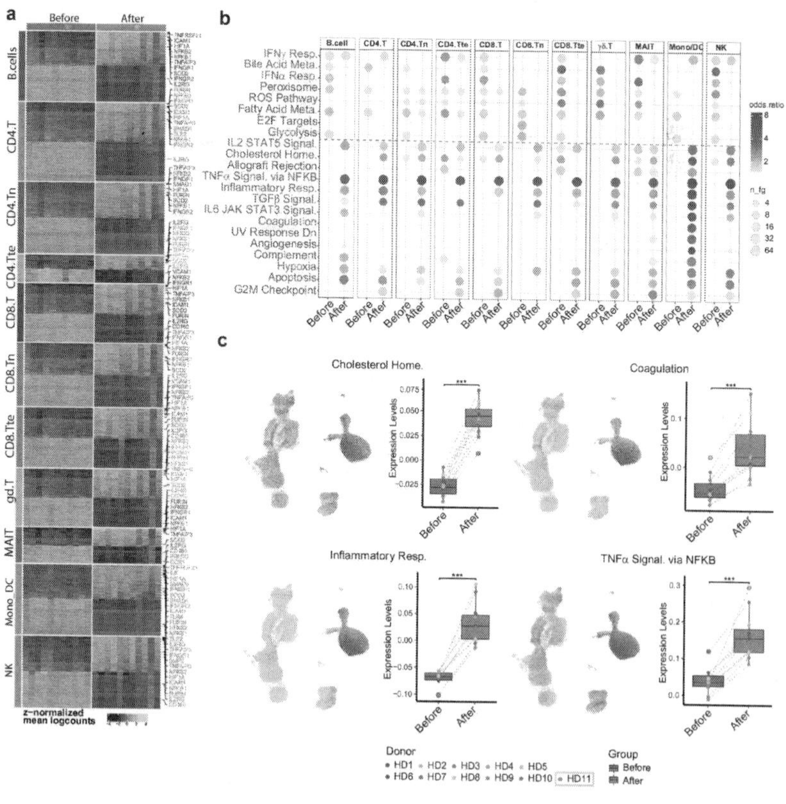

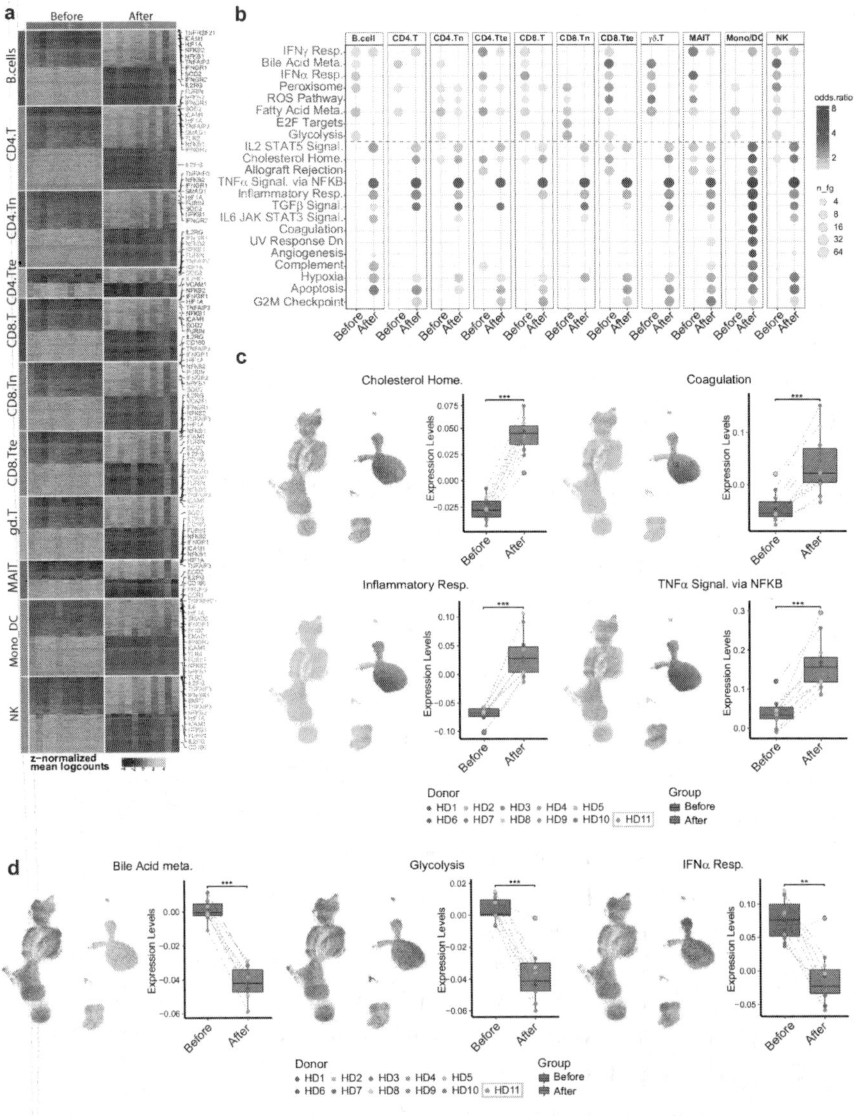

3.7 MECHANISTIC ASPECTS

Research by Wohhwa Lee and team (2020) published in *Signal Transduction and Targeted Therapy* probed the status of sterol regulatory element-binding protein-2 (SREBP-2) that is known to be activated by viral pathogens or cytokines (this becomes pertinent given the cytokine storm in COVID-19) in peripheral blood mononuclear cells (PBMCs) of COVID-19 patients. COVID-19 patients documented lower levels of total cholesterol (Ch), high-density lipoprotein (HDL)-Ch, and low-density lipoprotein (LDL)-Ch vs. controls with intensive care unit (ICU) patients demonstrative of lower levels of these factors than non-ICU patients. Deceased patients documented elevated SREBP-2 than those who survived with increased levels of cytokines associated with SREBP-2 like nuclear factor (NF)-κB, interleukin (IL)-1β and tumor necrosis factor (TNF)-α with increased disease severity. Subsequent inhibition of SREBP-2 with Fatostatin A and NF-κB signaling with SN50 diminished SREBP-2 activation in PBMCs of COVID-19 ICU patients along with inhibition of IL-1β and TNF-α. This pharmacological inhibition was also corroborated in cecal ligation and puncture (CLP) mice models with Fatostatin A and SN50 suppressing NF-κB signaling and SREBP-2 along with suppression of tissue damage markers.

Figure 22 below illustrates the involvement of sterol regulatory element-binding protein-2 (SREBP-2) in COVID-19 patients: ***TOP***: SREBP-2 in COVID-19: (a) SREBP-2 activation and COVID-19 severity (**$p < 0.01$). (b)SREBP-2 activation COVID-19 mortality (**$p < 0.01$). (c) NF-κB activation and COVID-19 severity (**$p < 0.01$). (d) NF-κB activation and patient survival (**$p < 0.01$). (e) IL-1β in COVID-19 (**$p < 0.01$). (f) TNF-α in patients (**$p < 0.01$). (g) Relative viability of PBMCs of COVID-19 patients vs. the *in vitro* culture time (**$p < 0.01$). (h) SREBP-2 activation in COVID-19 patient PBMCs (**$p < 0.01$, n.s. not significant) (Lee, W., Ahn, J.H., Park, H.H. et al, 2020: *Open Access: This article is licensed under a Creative Commons Attribution 4.0 International License, which permits use, sharing, adaptation, distribution and reproduction in any medium or format, as long as you give appropriate credit to the original author(s) and the source, provide a link to the Creative Commons license, and indicate if changes were made. The images or other third party material in this article are included in the article's Creative Commons license, unless indicated otherwise in a credit line to the material. If material is not included in the article's Creative Commons license and your intended use is not permitted by statutory regulation or exceeds the permitted use, you will need to obtain permission directly from the copyright holder. To view a copy of this license, visit http://creativecommons.org/licenses/by/4.0/*):

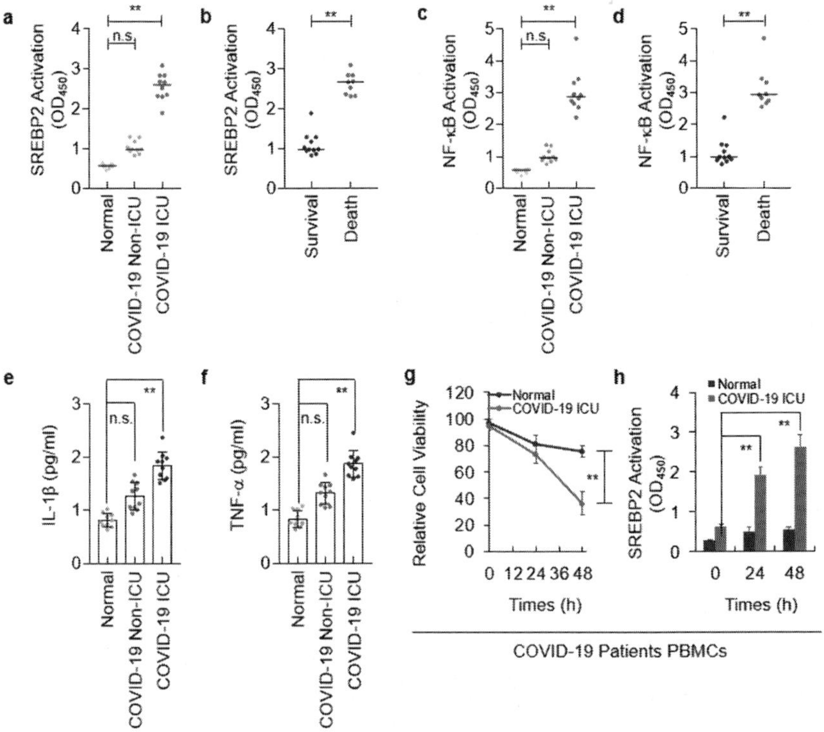

BOTTOM: Inhibiting SREBP-2 and NF-κB(a) SREBP-2 activation in PBMCs of COVID-19 ICU patients treated with NF-κB inhibitor (SN50) and SREBP-2 inhibitor (Fatostatin A) (**$p < 0.01$). IL-1β (b) and TNF-α (c) in PBMCs of COVID-19 ICU patients treated with SN50 and Fatostatin A (**$p < 0.01$). mRNA levels post-SN50 and Fatostatin A treatment: (d) SREBF2, (e) IL-1β, (f) TNF-α, (g) SCAP, (h) INSIG1, and (i) SIRT1 (**$p < 0.01$) (Lee, W., Ahn, J.H., Park, H.H. et al, 2020:

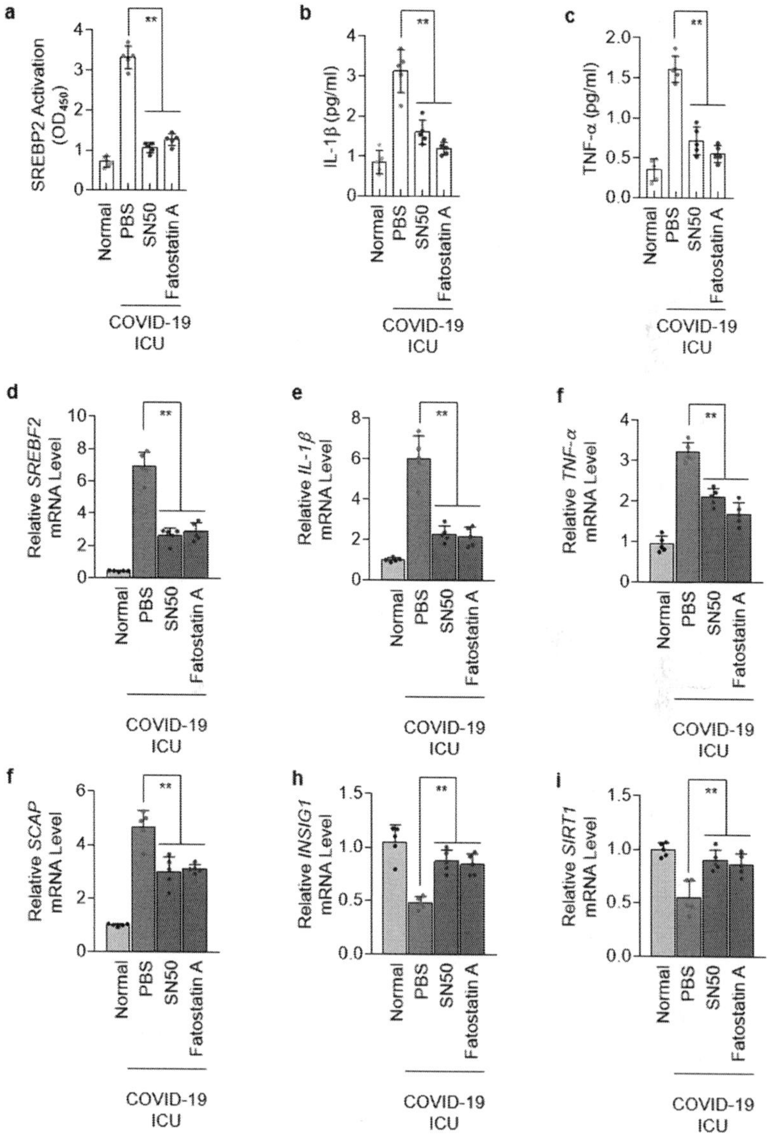
a
SREBP2 Activation (OD450)
b
IL-1β (pg/ml)
c
TNF-α (pg/ml)
d
Relative SREBF2 mRNA Level
e
Relative IL-1β mRNA Level
f
Relative TNF-α mRNA Level
f
Relative SCAP mRNA Level
h
Relative INSIG1 mRNA Level
i
Relative SIRT1 mRNA Level
Normal
PBS
SN50
Fatostatin A
COVID-19 ICU
**

Early research has documented the increased infiltration of immune cells linked to inflammation in atherosclerotic plaques (the formation of which entails cholesterol). Interestingly, the vital involvement of innate immune signaling complex, the NLRP3 inflammasome producing IL-1 cytokines in atherosclerosis has also been shown to be activated in patients who died of acute respiratory distress syndrome (ARDS) due to severe COVID-19. This has led to the plausible proposal that inflammation due to atherosclerosis induced by cholesterol crystals (CCs) and oxidized low-density lipoprotein (ox-LDL) set the stage for SARS-CoV-2. This can account for the dysregulated lipid metabolism in COVID-19 patients and the elevated occurrence of COVID-19 in those with cardiovascular disease (CVD).

Research has also outlined the involvement of cholesterol-rich microdomains facilitating the S protein spikes (S) - angiotensin-converting enzyme 2 (ACE2) receptor interactions. Notably, the cholesterol ester (CE) transporter HDL-scavenger receptor B type 1 (SR-B1) was shown to facilitate HCV (hepatitis C virus) entry and now for SARS-CoV-2 to allow its entry. Detailed elucidation of these mechanisms can offer a clearer picture.

(Tang et al, 2021)

Figure 23 below illustrates the role of cholesterol in COVID-19 infections A) ACE2 lipid rafts of the cell membrane is the SARS-CoV-2 receptor to enter the host cell while SR-B1 is the membrane receptor of HDL B) Increase in the membrane cholesterol increases the number of lipid rafts on the cell membrane and on the ACE2 receptor C) Increased blood cholesterol levels and CCs to facilitate plaque rupture to activate the NLRP3 inflammasome to form a favorable inflammatory environment for viral invasion; *Key: SARS-CoV-2: severe acute respiratory syndrome-coronavirus-2; ACE2: angiotensin-converting enzyme 2; SR-B1: scavenger receptor B type 1; HDL: high-density lipoprotein; ox-LDL: oxidized low-density lipoprotein; NLRP3: NLR family pyrin domain containing 3; IL-1β: interleukin-1 beta* (Tang et al, 2021: *This is an open access article distributed under the terms of the Creative Commons Attribution License (https://creativecommons.org/licenses/by/4.0/). See http://ivyspring.com/terms for full terms and conditions*):

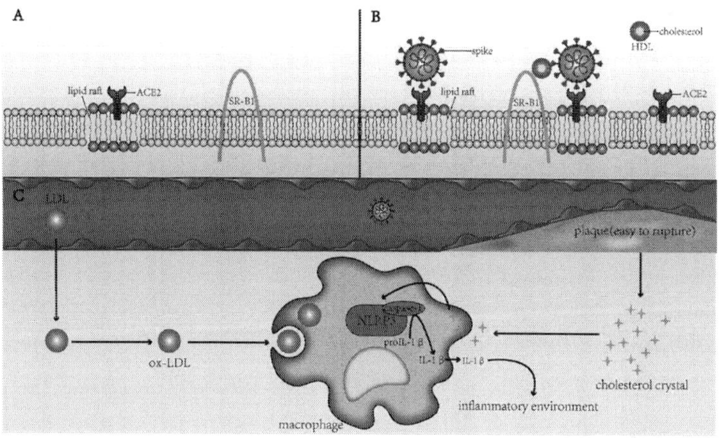

Looking at this more in detail, researchers Palacios-Rápalo and team (2021) summarized the cellular receptors on cholesterol-rich lipid rafts functioning in SARS-CoV-2 entry. This is documented in table 9 below ():

Receptor	Cellular function	Proposed entry mechanism	References
ACE-2	ACE-2 is a negative regulator of RAS and a catalyst for converting angiotensin II to angiotensin 1-7. ACE-2 is expressed in various organs such as the heart, lung, kidney, liver, etc.	ACE-2 binds to the S protein-RBD of SARS-CoV-2, facilitating virus entry via caveolin- or clathrin-dependent endocytosis.	(8)
TMPRSS2/4	TMPRSS are vital regulators of mammalian development and homeostasis in different tissues as the liver, lungs, pancreas, intestinal tract, and salivary glands.	TMPRSS2 and TMPRSS4 enhance cellular-virus membrane fusion by inducing protein S cleavage and exposing the fusion peptide, which interacts with the ACE-2 receptor	(13, 95)
HSPG	HSPG participates in multiple functions such as cellular adhesion and motility; moreover, they serve as receptors for endocytosis and are also involved in the control of numerous events that occur during inflammation	The interaction between SARS-CoV-2 spike protein and HSPG is necessary for the viral entry via endocytosis ACE-2-dependent.	(10, 96, 97)
Syndecan-1/4	Syndecans are expressed in various cellular sites and participate during adhesion between cell and extracellular matrix, cell-cell adhesion, cell migration, and regulation of the inflammatory response.	Syndecan-1/4 interacts with the S1 subunit of SARS-CoV-2 spike protein, an essential viral attachment factor, and mediator of viral entry.	(15, 98)
TLR-4	TLR4 is a key receptor that induces the pro-inflammatory response, can mediate inflammation by both exogenous and endogenous ligands, and is associated with chronic and acute diseases, promoting amplification of the inflammatory response.	TLR4 interacts with the S1 subunit of spike protein and is involved in SARS-CoV-2 entry, even if the cell line lacks the ACE-2 receptor. However, evidence on the mechanism of entry used by the virus is lacking.	(99-102)
CD147	CD147 is a transmembrane glycoprotein member of the immunoglobulin superfamily implicated in various physiological and pathological conditions due to its regulation of cell-cell recognition, cell differentiation, and tissue remodeling.	S protein of SARS-CoV-2 interacts with the CD147 receptor and facilitates virus entry via endocytosis even in the absence of the ACE-2 receptor.	(11, 103)
NRP1	NRP1 is a pleiotropic transmembrane polypeptide that acts as a growth factor or a cofactor in fibroblasts, platelets, hepatocytes, etc.	NRP1 enhances SARS-CoV-2 entry and infectivity only in co-expression with ACE-2 and TMPRSS2	(53)
L-SIGN	L-SIGN is a type II C-type lectin receptor involved in cell adhesion and pathogen recognition. It is expressed in dendritic cells, epithelial cells, lungs, liver, lymph nodes, and placenta	L-SIGN binds to high-mannose-type N-glycans present in the spike protein of SARS-CoV-2, favoring the viral entry in the presence of the ACE-2 receptor.	(16, 104)
AXL	AXL is a receptor tyrosine kinase; its activation promotes homodimerization, causing tyrosine autophosphorylation or phosphorylation of downstream targets, activating signaling pathways.	The NTD of the SARS-CoV-2 spike protein binds to AXL, independently of the presence of the ACE-2 receptor. However, low levels of the ACE-2 receptor synergize with the expression of the AXL to potentiate SARS-CoV-2 infection.	(14, 17, 105)
SR-B1	SR-B1 is the cell-surface HDL receptor that mediates a selective uptake system for cholesterol and other lipids in various cells, such as fibroblasts, hepatocytes, macrophages, adrenal, and alveolar cells.	The RBD of the SARS-CoV-2 S protein has an affinity for cholesterol and HDL components, enhancing the entry of the virus into the cell through SR-B1 only when ACE-2 es expressed.	(12, 106)

Let us see this in a diagrammatic representation with figure 24 below. The early phase entails the spike protein activation due to ACE-2 and TMPRSS2/4 facilitating virus-host cell fusion. Subsequent interaction of the ACE-2 receptor and a coreceptor like SR-B1, NRP1, HSPG, Syndecan-1/4 and L-SIGN for endocytic pathway activation (© 2021 Palacios-Rápalo, De Jesús-González, Cordero-Rivera, Farfan-Morales, Osuna-Ramos, Martínez-Mier, Quistián-Galván, Muñoz-Pérez, Bernal-Dolores, del Ángel and Reyes-Ruiz. *This is an open-access article distributed under the terms of the Creative Commons Attribution License (CC BY). The use, distribution or reproduction in other forums is permitted, provided the original author(s) and the copyright owner(s) are credited and that the original publication in this journal is cited, in accordance with accepted academic practice. No use, distribution or reproduction is permitted which does not comply with these terms*):

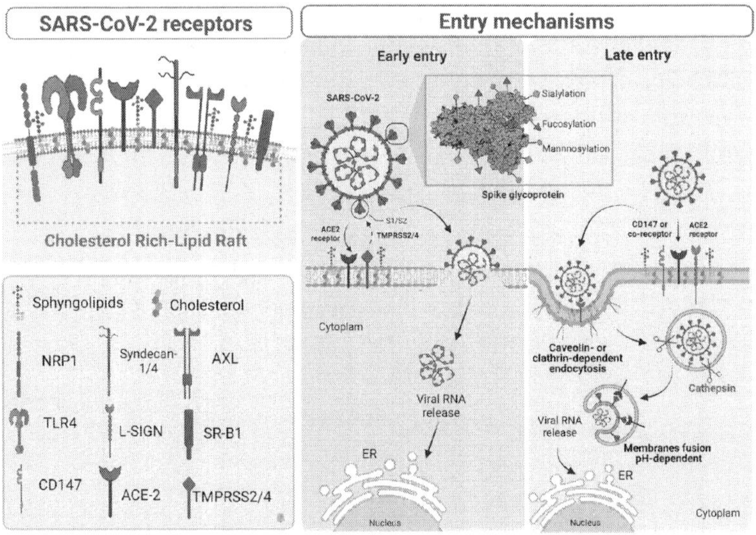

3.8 PLAUSIBLE CHOLESTEROL TARGETING FOR COVID-19 TREATMENT:

The maintenance of LDL-C values within the range is warranted for COVID-19 patients or healthy people. Serving as the predominant indicator for lipid lowering, normal levels of this molecule can lower the cardiovascular complications of COVID-19 while diminishing the risk to COVID-19. The lowering of lipids can entail the use of drugs, lifestyle management and lipid apheresis.

(Tang et al, 2021)

Table 10 below depicts the lipid values in a test (Tang et al, 2021: *This is an open access article distributed under the terms of the Creative Commons Attribution License (https://creativecommons.org/licenses/by/4.0/). See http://ivyspring.com/terms for full terms and conditions*):

Key: TC: total cholesterol; LDL-C: LDL cholesterol; HDL-C: HDL cholesterol; TG: Triglycerides

Category	U.S. and some other countries	Canada and most of Europe
TC	<200 mg/dL	<5.18 mmol/L
LDL-C	<100 mg/dL	<2.6 mmol/L
HDL-C	>60 mg/dL	>1.5 mmol/L
TG	<150 mg/dL	<1.7 mmol/L

Few drugs explored to target lipid rafts to stem coronavirus infections have been presented by Fecchi et al (2020) in table 11 below (© 2020 Fecchi, Anticoli, Peruzzu, Iessi, Gagliardi, Matarrese and Ruggieri. *This is an open-access article distributed under the terms of the Creative Commons Attribution License (CC BY). The use, distribution or reproduction in other forums is permitted, provided the original author(s) and the copyright owner(s) are credited and that the original publication in this journal is cited, in accordance with accepted academic practice. No use, distribution or reproduction is permitted which does not comply with these terms*):

Drugs	Mechanism of action	Target virus	Antiviral effects	References
MβCD, Nystatin	Disrupt lipid raft architecture by depleting cell membrane cholesterol	CRCoV HCoV-229E HCoV-NL63 HCoV-OC43 IBV MERS-CoV MHV SARS-CoV-2	Inhibit viral entry/endocytosis	Nomura et al., 2004; Thorp and Gallagher, 2004; Choi et al., 2005; Glende et al., 2008; Lu et al., 2008; Guo et al., 2017; Owczarek et al., 2018; Szczepanski et al., 2018; Baglivo et al., 2020; Fantini et al., 2020
Statins (mevastatin)	Inhibit lipid raft formation by lowering cholesterol biosynthesis	IBV SARS-CoV-2	Inhibit viral entry/endocytosis	Guo et al., 2017; Abu-Farha et al., 2020; Rodrigues-Diez et al., 2020
Chloroquine and Hydroxychloroquine	Interfere with lysosome-mediated autophagy function by increasing the endosomal/lysosomal pH	MERS-CoV HCoV-229E HCoV-OC43 SARS-CoV SARS-CoV-2	Prevent viral fusion and inhibit the viral entry by endocytosis, uncoating and exit (exocytosis) process	ClinicalTrials.gov, 2020a,b,c,d,e,f,g,h,i,j
Rapamycin (Sirolimus)	Inhibits mammalian target of rapamycin (mTOR) kinase	MERS-CoV PEDV SARS-CoV-2	Downregulates virus infection	ClinicalTrials.gov, 2020k
Nitazoxanide	Inhibits Akt/mTOR/ULK1 signaling pathway	HCoVOC43 MERS-CoV SARS-CoV-2	Inhibits virus replication	Shakya et al., 2018; Yang et al., 2020
Niclosamide	mTORC1 inhibitor	MERS-CoV SARS-CoV	Inhibits viral antigen synthesis	Liu et al., 2019; Xu et al., 2020

Preliminary work has unveiled the potential of FDA-approved cholesterol-lowering drugs for SARS-CoV-2 in table 12 below though further corroboration is warranted (© 2021 Palacios-Rápalo, De Jesús-González, Cordero-Rivera, Farfan-Morales, Osuna-Ramos, Martínez-Mier, Quistián-Galván, Muñoz-Pérez, Bernal-Dolores, del Ángel and Reyes-Ruiz. *This is*

Lipid-lowering Drug	Study type	Effect	References
ATV, RSV, SIM, PRV, FLV and PTV	Retrospective study: a. 13,981 patients diagnosed with SARS-Cov-2 in Hubei Province, China b. 2921 patients diagnosed with SARS-Cov-2, who are hospitalized in 150 Spanish hospitals.	Reduced risk of mortality among people with COVID-19	(188, 189)
MET	Retrospective studies: a. 283 diabetic patients hospitalized with confirmed SARS-Cov-2 in the Tongji Hospital of Wuhan, China. b. 1139 patients positive SARS-Cov-2 in 8 states in USA. c. 775 nursing Home Residents Infected with SARS-CoV2 from the Community Living Centers (CLC), USA.	a. Antidiabetic treatment with metformin was associated with lower hospitalization and mortality. b. Relative survival benefit in nursing home residents on metformin.	(204–206)
CQ/HCQ	Clinical study: a. Treating group of 100 COVID-19 patients treated with CQ. b. 36 patients diagnosed with SARS-Cov-2. Treatment group (20 patients) received HCQ 200 mg for ten days, three times a day (600 mg daily). Control group (16 patients). Six patients received AZI to prevent bacterial infections.	a. Improvements of pneumonia and lung imaging and reduction of the duration of illness without any adverse effects. b. On day six, treatment group showed a significant reduction in the viral load. The six patients who received a combination (HCQ and AZI) were testing negative, indicates the high effectiveness of the combination.	(207, 208)

SARS-CoV-2, severe acute respiratory syndrome coronavirus; STAs. Statins; ATV, Atorvastatin; RSV, Rosuvastatin; SIM, Simvastatin; PRV, Pravastatin; FLV, Fluvastatin; PTV, Pitavastatin; MET, Metformin; HCQ, Hydroxychloroquine; AZI, Azithromycin; CQ, Chloroquine.

A slew of studies has centered around the use of statins in COVID-19 patients. Surma, Banach & Lewek (2021) have summarized the results of meta-analyses probing the impact of statins on the course and prognosis of COVID-19 as depicted in table 13 below (Surma, S., Banach, M. & Lewek, J (2021):

Author/ year	Analyzed period	Number of in-cluded studies	Sample size	Results	Conclusions
Kow and Hasan [46]	to July 27, 2020	4	8990	Severity and mortality: risk ↓30% (HR = 0.70; 95% CI: 0.53–0.94)	The use of statins improves the prog-nosis in patients with COVID-19
Scheen [47]	Dec 2019 to Dec 2020	13	42,722	Mortality: no statistically risk reduction (OR = 0.97; 95% CI: 0.92–1.03)Severity: no statistically risk reduction (OR = 1.09; 95% CI: 0.99–1.22) Adjusted for confounders, a 27% reduction in the risk of severe disease and mortality in COVID-19 was demonstrated (adjusted OR = 0.73 ± 0.31 versus unadjusted OR = 1.44 ± 0.84; $p = 0.0028$).	Statin therapy in patients with CO-VID-19 may improve their prognosis
Onorato et al. [48]	2019 toSep 28, 2020	7	2398	Severity and mortality: risk ↓ by 41 (OR = 0.59; 95% CI: 0.35–0.99). The analysis of the results of studies in which statins were used before ad-mission to the hospital showed even greater benefits of their use (OR = 0.51; 95% CI: 0.41–0.64).	The use of statins improves the prog-nosis in patients with COVID-19
Pal et al. [49]	to Dec 18, 2020	14	19,988	Severity and mortality: no statistically risk reduction (OR = 1.02; 95% CI: 0.69–1.50). After adjustment reduce the risk of adverse outcomes by 49% (OR = 0.51; 95% CI: 0.41–0.63).	Statin therapy in patients with CO-VID-19 may improve their prognosis
Chow et al. [50]	Jan 2019 to Dec 2020	13	110,078	The use of statins before hos-pitalization: no significantly affect the risk of death (OR = 0.62; 95% CI: 0.38–1.03). The use of statins since the diag-nosis of COVID-19: reduced the risk of death (OR = 0.57; 95% CI: 0.43–0.75). The use of statins did not reduce the mortality of COVID-19 pa-tients admitted to the ICU (OR = 0.65; 95% CI: 0.26–1.64). Among patients in non-ICU, statin users were at lower risk of death (OR = 0.64; 95% CI: 0.46–0.88). The use of statins did not reduce the risk of admission to an ICU	Patients adminis-tered statins after COVID-19 diagnosis or non-ICU admitted patients were at lower risk of mortality

Vahedian-Azimi et al. [32]	to Nov 2, 2020	24	32,715	Significant reductions risk of ICU admission (OR = 0.78; 95% CI: 0.58–1.06). No significant effect on risk of tracheal intubation (OR = 0.79; 95% CI: 0.57–1.11). Significant reductions of death (OR = 0.70; 95% CI: 0.55–0.88). Was demonstrated that decrease mortality by in-hospital application of statins (OR = 0.40; 95% CI: 0.22–0.73), compared with pre-hospital use (OR = 0.77; 95% CI: 0.60–0.98)	Statins potentially reduction of ICU admission and total mortality reduction in COVID-19 patients
Wu et al [51].	to Nov 10, 2020	28	63,537	Statin use was associated with a reduction in the risk of mortality (OR = 0.71, 95% CI: 0.55–0.92) and the need for artificial ventilation (OR = 0.81, 95% CI: 0.69–0.95). Statin use was not found to reduce the risk of treatment in ICU (OR = 0.91; 95% CI: 0.55–1.51).	Statins can improve the prognosis of COVID-19 patients, so it does not seem necessary to stop taking them when the patient is admitted to the hospital.
Permana et al [52].	Dec 1, 2019 to Nov 11, 2020	13	52,122	In-hospital use of statin was associated with a decreased risk of mortality by 56% (RR = 0.54; 95% CI: 0.50–0.58). Pre-admission use of statin was not associated with risk of mortality (RR = 1.18; 95% CI: 0.79–1.77).	In-hospital use of statins was associated with a decreased risk of death in patients with COVID-19
Yetmar et al. [53]	to Dec, 2020	16	395,513	The use of statins before contracting COVID-19 reduced the risk of mortality (adjusted RR = 0.65; 95% CI: 0.56–0.77) and the risk of severe disease (aRR = 0.73; 95% CI: 0.57–0.94).	The use of statins is associated with a lower risk of death or serious illness in patients with COVID-19. The important role of the continued use of statins in patients indicated for lipid-lowering therapy during the COVID-19 pandemic is indicated.
Hariyanto and Kurniawan [54]	to Nov 25, 2020	35	11,930,583	Statin use has not significant effect on reducing the risk of COVID-19 (OR = 1.09; 95% CI: 0.58–2.03). Statin use has not significant effect on reducing the risk severity course of COVID-19 (OR = 1.07; 95% CI: 0.86–1.33).	Statin use did not improve the outcomes of patients with COVID-19.

Zein et al. [55]	to March 1, 2021	8	14,446	The use of statins reduced the risk of mortality (RR = 0.72; 95% CI: 0.55–0.95). In the subgroup of patients who used statins during hospitalization, an even lower risk of mortality was observed (RR = 0.71; 95% CI: 0.54–0.94). The observed effects were not influenced by such factors as: age, male gender, diabetes and arterial hypertension.	Statins reduce the risk of mortality in COVID-19 patients.
Diaz-Arocutipa et al. [56]	to March 3, 2021	25	147,824	In-hospital use of statins did not affect the risk of mortality (adjusted HR = 0.74; 95% CI: 0.49–1.10). Chronic statin use significantly reduced the risk of mortality (aHR = 0.71; 95% CI: 0.56–0.91).	Statins, especially when used chronically, reduce the risk of death in patients with COVID-19
Kollias et al. [57]	to March 5, 2021	22	114,688	Using statins versus not using statins reduced the risk of mortality (HR = 0.65; 95% CI: 0.53–0.81 and OR = 0.65; 95% CI: 0.55–0.78).	Statin treatment was associated with an approximately 35% reduction in adjusted risk of death in hospitalized COVID-19 patients.
Kow and Hasan [58]	to Jun 3, 2021	35	138,402	Statin use reduced the risk of mortality from any cause (OR = 0.63, 95% CI: 0.51–0.79), and the risk of severe COVID-19 (OR = 0.80, 95% CI: 0.73–0.88).	The use of statins is associated with a better prognosis in COVID-19 patients.

The various impacts of statins on COVID-19 are also summarized by researchers Surma, Banach & Lewek (2021) as depicted in figure 25 below (Surma, S., Banach, M. & Lewek, J (2021): *Open Access: This article is licensed under a Creative Commons Attribution 4.0 International License, which permits use, sharing, adaptation, distribution and reproduction in any medium or format, as long as you give appropriate credit to the original author(s) and the source, provide a link to the Creative Commons licence, and indicate if changes were made. The images or other third party material in this article are included in the article's Creative Commons licence, unless indicated otherwise in a credit line to the material. If material is not included in the article's Creative Commons licence and your intended use is not permitted by statutory regulation or exceeds the permitted use, you will need to obtain permission directly from the copyright holder. To view a copy of this licence, visit http://creativecommons.org/licenses/by/4.0/. The Creative Commons Public Domain Dedication waiver (http://creativecommons.org/*

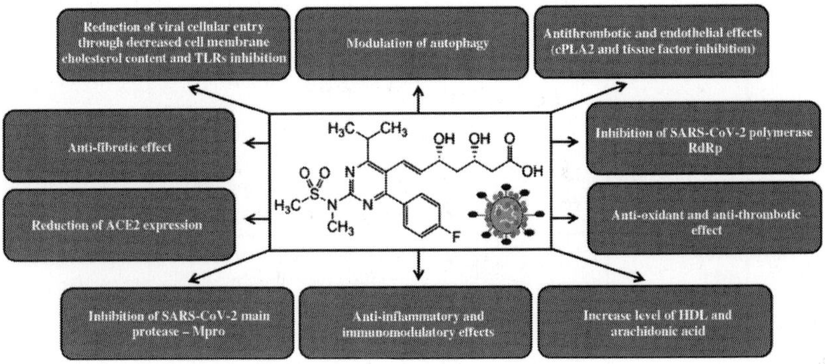

3.9 FINAL POINTS

Research has unveiled that lipid disorders caused in COVID-19 are on the same lines as that of other inflammatory conditions and infections. While studies have pointed out hypolipidemia, further research is warranted before empirical conclusions. Figure 26 below summarizes the lipid disorders caused due to SARS-CoV-2 - severe acute respiratory syndrome coronavirus 2 as depicted excellently by Surma and team (2021) : *Key: TMPRSS2 - transmembrane serine protease 2; ACE2 - angiotensin-converting enzyme 2; RNA - ribonucleic acid; IL-6 – interleukin 6; MCP1 - monocyte chemoattractant protein-1; MIP - macrophage inflammatory proteins; Apo CIII - apolipoprotein C-III; Apo A1 – apolipoprotein A1; Apo E – apolipoprotein E; SAA - serum amyloid A; TC – total cholesterol; HDL – high-density lipoprotein; LDL – low-density lipoprotein; VLDL – very-low-density lipoprotein; TG – triglyceride* (Surma, S., Banach, M. & Lewek, J (2021):

creativecommons.org/licenses/by/4.0/. The Creative Commons Public Domain Dedication waiver (http://creativecommons.org/publicdomain/zero/1.0/) applies to the data made available in this article, unless otherwise stated in a credit line to the data):

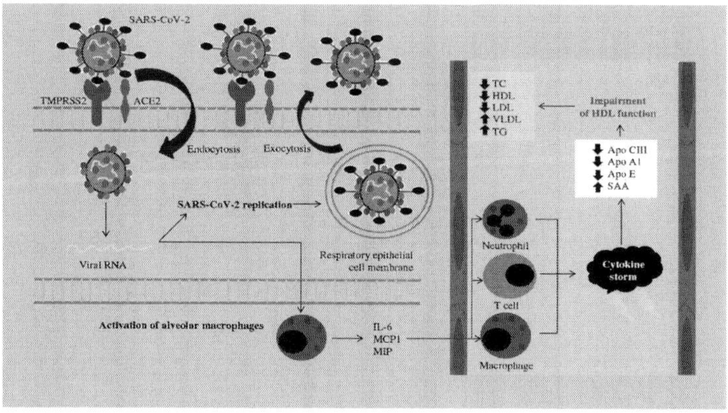

Researchers Li and team (2021) published a 3–6 month-follow-up study in *BMC Infectious Diseases* of ALT, ALP, GGT, LDL-c, high-density lipoprotein cholesterol (HDL-c), total cholesterol (TC), triglycerides (TG), total protein (TP), albumin (ALB), globulin (GLB), total bilirubin (TBIL), direct bilirubin (DBIL), and total bile acid (TBA) of discharged COVID-19 patients (107-mild = 59 and severe/critical = 48). For mild cases, LDL-c levels documented an evident increase at the follow-up vs. at the time of admission while LDL-c and HDL-c levels were more at follow-up than at admission. Table 14 below documents the changes in cholesterol and other lab factors in COVID-19 patients at follow-up post-discharge (Li, G., Du, L., Cao, X. et al, 2021: *Open Access: This article is licensed under a Creative Commons Attribution 4.0 International License, which permits use, sharing, adaptation, distribution and reproduction in any medium or format, as long as you give appropriate credit to the original author(s) and the source, provide a link to the Creative Commons licence, and indicate if changes were made. The images or other third party material in this article are included in the article's Creative Commons licence, unless indicated otherwise in a credit line to the material. If material is not included in the article's Creative Commons licence and your intended use is not permitted by statutory regulation or exceeds the permitted use, you will need to obtain permission directly from the copyright holder. To view a copy of this licence, visit http://creativecommons.org/licenses/by/4.0/. The Creative Commons Public Domain Dedication waiver (http://creativecommons.org/*

publicdomain/zero/1.0/) applies to the data made available in this article, unless otherwise stated in a credit line to the data):

Key: Wilcoxon signed-rank test (two-sided)employed for assessment; Adm at time of admission, *Flw* follow-up, *ATIII antithrombin III, FDP* fibrin degradation product, *FIB* fibrinogen, *ALT* alanine aminotransferase, *ALP* alkaline phosphatase, *GGT* gamma-*glutamyl transferase, WBC* white blood cell, *LY* lymphocyte, *MO* monocyte, *TC* total cholesterol, *TG* triglycerides, *TP* total protein, *ALB* albumin, *GLB* globulin, *TBIL* total bilirubin, *DBIL* direct bilirubin, *CRP* C-reactive protein, *TBA* total bile acid, *A/G* ALB/GLB ratio

Laboratory values	**Reference ranges**	**COVID-19 follow-up patients**					
		Mild (*n* = 59)			**Severe/Critical (*n* = 48)**		
		Adm	**Flw**	***P***	**Adm**	**Flw**	***p***
LDL-c	*109 (29) mg/dL	100.5 (87.0, 113.3)	103.6 (90.5, 116.0)	***0.048***	96.1 (74.3, 107.2)	103.6 (76.2, 123.4)	***0.003***
HDL-c	*52 (15) mg/dL	52.9 (46.0, 64.2)	54.5 (46.8, 63.4)	0.297	50.3 (43.7, 65.4)	55.3 (46.0, 65.7)	***0.042***
TC	*182 (34) mg/dL	197.9 (170.5, 223.5)	203.4 (187.2, 220.8)	***0.042***	192.4 (151.3, 221.6)	203.2 (152.1, 246.6)	***0.016***
Non-HDL-c	*130 (35) mg/dL	144.2 (118.3, 160.8)	147.0 (125.3, 165.9)	***0.039***	135.7 (102.8, 157.4)	151.8 (103.7, 182.4)	**0.009**
TG	*133 (99) mg/dL	165.5 (131.1, 219.6)	140.8 (108.1, 177.1)	***0.001***	141.7 (107.6, 182.2)	143.9 (111.6, 200.6)	0.376
WBC	3.5–9.5 (×10^9/L)	5.8 (5.2, 6.8)	5.4 (4.7, 6.2)	***0.002***	6.3 (5.0, 8.1)	5.6 (5.0, 6.7)	***0.046***
LY	*2.1 (0.5) (× 10^9/L)	1.5 (1.3, 1.8)	1.9 (1.5, 2.2)	< ***0.001***	1.6 (1.1, 1.9)	2.0 (1.5, 2.3)	< ***0.001***
MO	0.1–0.6 (×10^9/L)	0.5 (0.3, 0.6)	0.3 (0.2, 0.4)	< ***0.001***	0.5 (0.3, 0.7)	0.3 (0.3, 0.4)	< ***0.001***
D-dimer	< 0.5 mg/ml	0.5 (0.2, 0.8)	0.3 (0.2, 0.4)	< ***0.001***	0.7 (0.4, 1.1)	0.4 (0.3, 0.5)	< ***0.001***
ATIII	80–120%	90 (81.5, 96.5)	97 (89, 102.5)	< ***0.001***	88 (81.2, 93.8)	96 (87, 102.8)	***0.001***
FDP	< 5 μg/L	1.7 (1.1, 2.5)	1.1 (1, 1.5)	< ***0.001***	2.7 (1.6, 3.7)	1.3 (1.0, 2.0)	< ***0.001***
FIB	2–4 g/L	4.0 (3.4, 5.2)	3.2 (2.7, 3.5)	< ***0.001***	4.6 (3.7, 5.5)	3.3 (3.0, 3.7)	< ***0.001***

CRP	< 4 mg/L	1.7 (0.6, 4.8)	1.4 (0.6, 2.2)	< ***0.001***	2.9 (1.2, 8.3)	1.6 (1.1, 2.6)	***0.002***
ALT	5–35 U/L	36 (22, 54)	25 (19, 32)	< ***0.001***	33.5 (21.8, 62.2)	21 (15, 32.4)	***0.013***
ALP	40–150 U/L	89 (72, 104)	77 (64, 87)	< ***0.001***	91 (71, 107)	73 (64, 85)	< ***0.001***
GGT	7–32 U/L	32 (22, 51)	23 (17, 32)	< ***0.001***	31.5 (18.7, 50.5)	23 (18, 31)	< ***0.001***
TBIL	5.1–19 μmol/L	11.5 (9.8, 14.6)	16.7 (12.2, 20.8)	< ***0.001***	11.1 (8.8, 14.3)	13.2 (10.8, 15.9)	***0.002***
DBIL	1.7–6.8 μmol/L	3.1 (2.5, 3.6)	3.7 (3, 4.5)	< ***0.001***	3.2 (2.4, 4.0)	3.2 (3, 3.7)	0.215
TBA	0–10 μmol/L	4.3 (3.2, 6.9)	2.5 (1.5, 3.7)	< ***0.001***	4.7 (3.3, 7.7)	2.6 (1.3, 4.6)	< ***0.001***
TP	64–83 g/L	70.6 (67.6, 74.7)	73.8 (71.9, 77.2)	< ***0.001***	68.9 (65.4, 72.3)	75.4 (72.8, 79.5)	< ***0.001***
GLB	20–30 g/L	28.6 (26.4, 31.7)	27.4 (24.7, 29.7)	***0.002***	28.9 (26.4, 32.6)	29.0 (25.5, 31.8)	0.506
ALB	35–55 g/L	41.9 (39.4, 44)	46.8 (45.7, 49)	< ***0.001***	37.4 (6.4)	46.5 (2.7)	< ***0.001***
A/G ratio	1.5–2.5	1.5 (1.3, 1.6)	1. 7 (1.5, 1.9)	< ***0.001***	1.3 (1.2, 1.5)	1.6 (1.5, 1.8)	< ***0.001***

Such analyses are indicative of the "long-term" recovery entailed post-COVID-19 infections.

The impact of COVID-19 is multipronged and here the changes in the cholesterol profile as outlined by research have been presented. Many reports have unveiled the lowering in HDL (vitally involved in targeting inflammation and toxins) in COVID-19. Changes that diminish the protective functioning of HDL i.e. serum amyloid A-1 (SAA1) replacing ApoA1, elevated core triglycerides and lowered ApoE, ApoM and paraoxonase 1 (PON1) have been documented in inflammatory/cardiometabolic disease and COVID-19(Kluck GEG, Yoo J-A, Sakarya EH, Trigatti BL, 2021).

Figure 27 below illustrates the remodeling of HDL in inflammatory/cardiometabolic disease and COVID-19 (Kluck GEG, Yoo J-A, Sakarya EH, Trigatti BL, 2021: © 2021 by the authors. Licensee MDPI, Basel, Switzerland. *This article is an open access article distributed under the terms and conditions of the Creative Commons Attribution (CC BY) license: https://creativecommons.org/licenses/by/4.0/*):

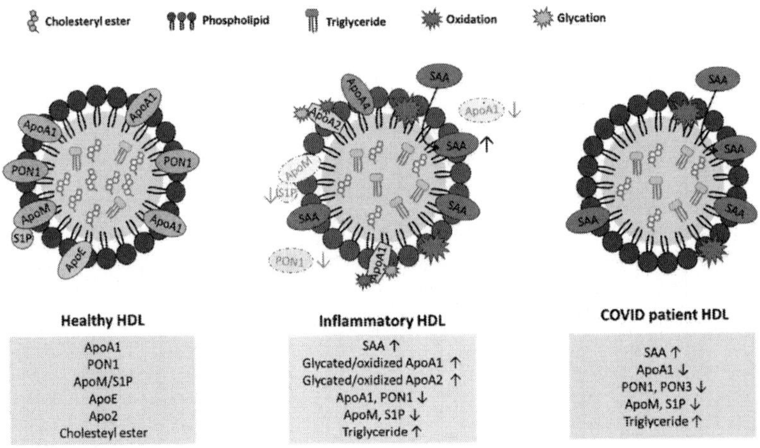

Table 15 below documents alterations in HDL in COVID-19 patients as documented in studies :

Author	Country	Number of patients	Time point	Comparison of HDL levels
Wang (Wang et al. (2020a)	China	228	Within 24 h after admission	COVID-19 patients vs healthy control: median, 0.78 vs 1.37 mmol/L, $p < 0.001$ Severe vs non-severe patients: median, 0.69 vs 0.79 mmol/L, $p = 0.032$
Huang (Huang et al. (2021)	China	218	The 1st day of admission	COVID-19 patients vs healthy control: mean, 1.02 vs 1.52 mmol/L, $p < 0.05$ Severe vs non-severe patients: mean, 0.83 vs 1.15 mmol/L, $p < 0.05$
Zhang (Zhang B. et al. (2020)	China	74	Not known	Severe vs non-severe patients: median, 0.92 vs 1.08 mmol/L, $p = 0.021$
Xie (Xie et al. (2020)	China	62	Not known	Severe vs non-severe patients with CVD: median, 1.1 vs 1.4 mmol/L Severe vs non-severe patients without CVD: median, 1.1 vs 1.3 mmol/L
Hu (Hu et al. (2020)	China	114	On admission	COVID-19 patients vs healthy control: mean, 1.08 vs 1.27 mmol/L, $p < 0.001$ Severe vs non-severe patients: median, 1.01 vs 1.21 mmol/L, $p < 0.001$
Wei (Wei et al. (2020b)	China	597	Not known	Mild vs severe vs critical patients: median, 50 vs 50 vs 36 mg/dL, $p < 0.05$
Tanaka (Tanaka et al. (2020a)	France	48	Upon ICU admission	Alive vs dead patients on day 28: median, 0.6 vs 0.5 mmol/L, $p = 0.036$
Huang (Huang W. et al. (2020)	China	2,623	At admission	Critical vs non-critical patients: median, 0.86 vs 0.95 mmol/L, $p < 0.001$
Sun (Sun et al. (2020)	China	99	Within 24 h of admission	Mild vs severe: median, 1.18 vs 0.94, $p < 0.001$
Ouyang (Ouyang et al. (2020)	China	107	Last result	Survivors vs non-survivors: average, 1.07 vs 0.79 mmol/L, $p = 0.006$
Li (Li Y. et al. (2021)	China	424	Not known	Survivors vs non-survivors: median, 0.9 vs 0.8, $p = 0.001$
Turgay (Turgay Yildirim and Kaya (2021)	Turkey	139	At admission	Survivors vs non-survivors: median, 44.0 vs 28.5 mg/dL, $p < 0.001$

Abbreviations: HDL, high-density lipoprotein; COVID-19, Coronavirus disease 2019.

Research is ongoing to address the altered profile of cholesterol given that COVID-19 is yet to see a "standard treatment". The answer lies in continued and detailed research on these lines.

"Listen to the mustn'ts, child. Listen to the don'ts. Listen to the shouldn'ts, the impossibles, the won'ts. Listen to the never haves, then listen close to me ... Anything can happen, child. Anything can be." —
Shel Silverstein.

3.10 REFERENCES

1. Knoops K, Kikkert M, Worm SHEvd, Zevenhoven-Dobbe JC, van der Meer Y, Koster AJ, et al. (2008) SARS-Coronavirus Replication Is Supported by a Reticulovesicular Network of Modified Endoplasmic Reticulum. PLoS Biol 6(9): e226. https://doi.org/10.1371/journal.pbio.0060226
2. Nagy PD, Strating JRPM, van Kuppeveld FJM (2016) Building Viral Replication Organelles: Close Encounters of the Membrane Types. PLoS Pathog 12(10): e1005912. https://doi.org/10.1371/journal.ppat.1005912
3. Tanaka S, De Tymowski C, Assadi M, Zappella N, Jean-Baptiste S, Robert T, et al. (2020) Lipoprotein concentrations over time in the intensive care unit COVID-19 patients: Results from the ApoCOVID study. PLoS ONE 15(9): e0239573. https://doi.org/10.1371/journal.pone.0239573
4. Dias SSG, Soares VC, Ferreira AC, Sacramento CQ, Fintelman-Rodrigues N, Temerozo JR, et al. (2020) Lipid droplets fuel SARS-CoV-2 replication and production of inflammatory mediators. PLoS Pathog 16(12): e1009127. https://doi.org/10.1371/journal.ppat.1009127
5. Wang, R., Simoneau, C. R., Kulsuptrakul, J., Bouhaddou, M., Travisano, K. A., Hayashi, J. M., Carlson-Stevermer, J., Zengel, J. R., Richards, C. M., Fozouni, P., Oki, J., Rodriguez, L., Joehnk, B., Walcott, K., Holden, K., Sil, A., Carette, J. E., Krogan, N. J., Ott, M., & Puschnik, A. S. (2021). Genetic Screens Identify Host Factors for SARS-CoV-2 and Common Cold Coronaviruses. *Cell*, *184*(1), 106–119.e14. https://doi.org/10.1016/j.cell.2020.12.004
6. Sarmadi, M., Ahmadi-Soleimani, S.M., Fararouei, M. et al. COVID-19, body mass index and cholesterol: an ecological study using global data. BMC Public Health 21, 1712 (2021). https://doi.org/10.1186/s12889-021-11715-7
7. Wang, G., Zhang, Q., Zhao, X. et al. Low high-density lipoprotein level is correlated with the severity of COVID-19 patients: an observational study. Lipids Health Dis 19, 204 (2020). https://doi.org/10.1186/s12944-020-01382-9
8. Masana, L., Correig, E., Ibarretxe, D. et al. Low HDL and high triglycerides predict COVID-19 severity. Sci Rep 11, 7217 (2021). https://doi.org/10.1038/s41598-021-86747-5

9. Begue, F., Tanaka, S., Mouktadi, Z. et al. Altered high-density lipoprotein composition and functions during severe COVID-19. Sci Rep 11, 2291 (2021). https://doi.org/10.1038/s41598-021-81638-1
10. Vlasov I, Panteleeva A, Usenko T, Nikolaev M, Izumchenko A, Gavrilova E, Shlyk I, Miroshnikova V, Shadrina M, Polushin Y, Pchelina S, Slonimsky P. Transcriptomic Profiles Reveal Downregulation of Low-Density Lipoprotein Particle Receptor Pathway Activity in Patients Surviving Severe COVID-19. *Cells*. 2021; 10(12):3495. https://doi.org/10.3390/cells10123495
11. Caterino, M., Gelzo, M., Sol, S. et al. Dysregulation of lipid metabolism and pathological inflammation in patients with COVID-19. Sci Rep 11, 2941 (2021). https://doi.org/10.1038/s41598-021-82426-7
12. Meoni, G., Ghini, V., Maggi, L., Vignoli, A., Mazzoni, A., Salvati, L., Capone, M., Vanni, A., Tenori, L., Fontanari, P., Lavorini, F., Peris, A., Bartoloni, A., Liotta, F., Cosmi, L., Luchinat, C., Annunziato, F., & Turano, P. (2021). Metabolomic/lipidomic profiling of COVID-19 and individual response to tocilizumab. PLoS pathogens, 17(2), e1009243. https://doi.org/10.1371/journal.ppat.1009243
13. Schmelter F, Föh B, Mallagaray A, Rahmöller J, Ehlers M, Lehrian S, von Kopylow V, Künsting I, Lixenfeld AS, Martin E, Ragab M, Meyer-Saraei R, Kreutzmann F, Eitel I, Taube S, Käding N, Jantzen E, Graf T, Sina C and Günther UL (2021) Metabolic and Lipidomic Markers Differentiate COVID-19 From Non-Hospitalized and Other Intensive Care Patients. Front. Mol. Biosci. 8:737039. doi: 10.3389/fmolb.2021.737039
14. Li, Y., Hou, G., Zhou, H. et al. Multi-platform omics analysis reveals molecular signature for COVID-19 pathogenesis, prognosis and drug target discovery. Sig Transduct Target Ther 6, 155 (2021). https://doi.org/10.1038/s41392-021-00508-4
15. Ballout RA, Kong H, Sampson M, Otvos JD, Cox AL, Agbor-Enoh S, Remaley AT. The NIH Lipo-COVID Study: A Pilot NMR Investigation of Lipoprotein Subfractions and Other Metabolites in Patients with Severe COVID-19. Biomedicines. 2021; 9(9):1090. https://doi.org/10.3390/biomedicines9091090
16. Liu, J., Wang, J., Xu, J. *et al.* Comprehensive investigations revealed consistent pathophysiological alterations after vaccination with COVID-19 vaccines. *Cell Discov* 7, 99 (2021).https://doi.org/10.1038/s41421-021-00329-3

17. Lee, W., Ahn, J.H., Park, H.H. et al. COVID-19-activated SREBP2 disturbs cholesterol biosynthesis and leads to cytokine storm. Sig Transduct Target Ther 5, 186 (2020). https://doi.org/10.1038/s41392-020-00292-7
18. Tang Y, Hu L, Liu Y, Zhou B, Qin X, Ye J, Shen M, Wu Z, Zhang P. Possible mechanisms of cholesterol elevation aggravating COVID-19. *Int J Med Sci* 2021; 18(15):3533-3543. doi:10.7150/ijms.62021. Available from https://www.medsci.org/v18p3533.htm
19. Palacios-Rápalo SN, De Jesús-González LA, Cordero-Rivera CD, Farfan-Morales CN, Osuna-Ramos JF, Martínez-Mier G, Quistián-Galván J, Muñoz-Pérez A, Bernal-Dolores V, del Ángel RM and Reyes-Ruiz JM (2021) Cholesterol-Rich Lipid Rafts as Platforms for SARS-CoV-2 Entry. *Front. Immunol.* 12:796855. doi: 10.3389/fimmu.2021.796855
20. Fecchi K, Anticoli S, Peruzzu D, Iessi E, Gagliardi MC, Matarrese P and Ruggieri A (2020) Coronavirus Interplay With Lipid Rafts and Autophagy Unveils Promising Therapeutic Targets. *Front. Microbiol.* 11:1821. doi: 10.3389/fmicb.2020.01821
21. Surma, S., Banach, M. & Lewek, J. COVID-19 and lipids. The role of lipid disorders and statin use in the prognosis of patients with SARS-CoV-2 infection. Lipids Health Dis 20, 141 (2021). https://doi.org/10.1186/s12944-021-01563-0
22. Li, G., Du, L., Cao, X. et al. Follow-up study on serum cholesterol profiles and potential sequelae in recovered COVID-19 patients. BMC Infect Dis 21, 299 (2021). https://doi.org/10.1186/s12879-021-05984-1
23. Kluck GEG, Yoo J-A, Sakarya EH, Trigatti BL. Good Cholesterol Gone Bad? HDL and COVID-19. International Journal of Molecular Sciences. 2021; 22(19):10182. https://doi.org/10.3390/ijms221910182
24. Wang G, Deng J, Li J, Wu C, Dong H, Wu S and Zhong Y (2021) The Role of High-Density Lipoprotein in COVID-19. Front. Pharmacol. 12:720283. doi: 10.3389/fphar.2021.720283

CHAPTER 4

DNA, Gene Interactions and Gender: Research Snapshots

Contents

4.1 IT IS IN THE GENES AND THEIR INTERACTIONS

Research is unveiling the involvement of genes and the environment for blood lipids like cholesterol (Kathiresan et al, 2007). Bennet and team (2007) probed the associations of apolipoprotein E (apoE) genotype in lipid levels with coronary risk. Analyses of 82 studies (86067 healthy participants) and 121 studies of coronary outcomes (37850 cases and 82727 controls) unveiled a lowering of mean low-density lipoprotein cholesterol (LDL-C) values by approximately 31% in the *ε2/ε2* genotype vs. that of the *ε4/ε4* genotype. An inverse relationship emerged for *ε2/ε2*, *ε2/ε3*, *ε2/ε4*, *ε3/ε3*, *ε3/ε4*, *ε4/ε4* genotypes with LDL-C and with coronary risk. The risk of coronary heart disease was 20% lower in *ε2* carriers vs. the *ε3/ε3* genotype while the risk for *ε4/ε4* genotypes is more (Bennet et al, 2007).

The field of genomics has facilitated the scrutiny of the impact of genotype variation on phenotypic variation in a genome-wide association study (GWAS). This approach has allowed its use to unveil previously unknown genetic loci involved in a slew of maladies like inflammatory bowel disease, macular degeneration, etc.

Kathiresan and team (2007) reported genome-wide analyses (Affymetrix 100K GeneChip) for fasting blood lipid traits in Framingham Heart Study Offspring cohort participants (mean age=47 years; n=1087; 52% women). GWAS entailed generalized estimating equations (GEE), family-based association tests (FBAT) and variance components linkage. The phenotypes scored are depicted in table 1 below (Kathiresan S, Manning AK, Demissie S et al, 2007: Open Access: This article is published under license to BioMed Central Ltd. *This is an Open Access article is distributed under the terms of the Creative Commons Attribution License (https://creativecommons.org/licenses/by/2.0), which permits unrestricted use, distribution, and reproduction in any medium, provided the original work is properly cited*):

Phenotype	Acronym	Phenotypes	N	h^2 *	Offspring Exam	Adjustment† Multivariable model
Total cholesterol	TC	7	1069	0.57‡	1,2,3,4,5,6,7	Age, age^2, smoking, body mass index, alcohol consumption, menopausal status, hormone replacement therapy; Covariate-adjusted residuals created separately by gender
Low-density lipoprotein cholesterol	LDL-C	7	1056	0.59‡	1,2,3,4,5,6,7	
High-density lipoprotein cholesterol	HDL-C	7	1062	0.52‡	1,2,3,4,5,6,7	
Triglycerides	TG	7	1068	0.48‡	1,2,3,4,5,6,7	
Mean total cholesterol	MeanTC	1	1087	0.61	Avg of Exams 1,2,3,4,5,6,7	
Mean low-density lipoprotein cholesterol	Mean-LDL-C	1	1086	0.66	Avg of Exams 1,2,3,4,5,6,7	
Mean high-density lipoprotein cholesterol	Mean-HDL-C	1	1087	0.69	Avg of Exams 1,2,3,4,5,6,7	
Mean triglycerides	MeanTG	1	1087	0.58	Avg of Exams 1,2,3,4,5,6,7	
Apolipoprotein A-I	ApoA-I	1	997	0.42	4	
Apolipoprotein B	ApoB	1	997	0.47	4	
Apolipoprotein C3	ApoC3	1	767	0.38	4	
Apolipoprotein E	ApoE	1	744	0.54	5	
High-density lipoprotein 2 cholesterol	HDL2-C	2	955 (exam 4) 958 (exam 5)	0.50 0.53	4,5	

High-density lipoprotein 3 cholesterol	HDL3-C	2	984 (exam 4) 1020 (exam 5)	0.48 0.54	4,5	
Large high-density lipoprotein by NMR	HDL-NMRlg	1	851	0.53	4	
Intermediate high-density lipoprotein by NMR	HDL-NMRint	1	851	0.22	4	
Small high-density lipoprotein by NMR	HDL-NMRsm	1	851	0.24	4	
High-density lipoprotein particle size by NMR	HDL-NMRsz	1	851	0.50	4	
Intermediate-density lipo-protein particle by NMR	IDLNMR	1	851	0.20	4	
Large low-density lipoprotein by NMR	LDL-NMRlg	1	851	0.39	4	
Small low-density lipoprotein by NMR	LDL-NMRsm	1	851	0.40	4	
Low-density lipoprotein particle size by NMR	LDL-NMRsz	1	851	0.40	4	
Large very low-density lipoprotein particle by NMR	VLDL-NMRlg	1	851	0.34	4	
Intermediate very low-density lipoprotein particle by NMR	VLDL-NMRint	1	878	0.40	4	

Small very low-density lipoprotein particle by NMR	VLDL-NMRsm	1	851	0.24	4	
Very low-density lipoprotein particle size by NMR	VLDL-NMRsz	1	851	0.42	4	
Triglyceride/high-density lipoprotein cholesterol ratio	tghdl	7	1060	0.56‡	1,2,3,4,5,6,7	
Total cholesterol/high-density lipoprotein cholesterol ratio	cholhdl	7	1060	0.58‡	1,2,3,4,5,6,7	
Lipoprotein(a)	Lp(a)	1	763	0.90	3	
Remnant lipoprotein cholesterol	RLP-C	1	746	0.34	4	
Remnant lipoprotein triglycerides	RLP-TG	1	715	0.33	4	

The long-term averages of LDL-C, HDL-C, and TG (seven measurements) over a timeframe of close to 30 years unveiled the highly heritable character of LDL-C, HDL-C, and TG (h^2 = 0.66, 0.69, 0.58, respectively; $P < 0.0001$). Several candidate genes were illuminated that were on the lines of earlier reports. Table 2 below depicts the outcome of the GWAS analyses as reported by the scientists (Kathiresan S, Manning AK, Demissie S et al, 2007: Open Access: This article is published under license to BioMed Central Ltd. *This is an Open Access article is distributed under the terms of the Creative Commons Attribution License (https://creativecommons.org/licenses/by/2.0), which permits unrestricted use, distribution, and reproduction in any medium, provided the original work is properly cited*):

Gene	Phenotype	Selected SNPs previously associated with phenotype	# SNPs in Affymetrix 100K within 60 kb of gene	# SNPs in Affymetrix 100K within 60 kb of Gene Locus and with $r^2 > 0.5$ to previously reported SNP	SNP in Affymetrix 100K (r^2 to previously reported SNP)	p for association in FHS 100K
APOE	LDL-C	rs429358 rs7412	1	0	-	-
PCSK9	LDL-C	rs11591147	7	0	-	-
CETP	HDL-C	rs1800775	2	0	-	-
LIPC	HDL-C	rs1800588	18	0	-	-
LPL	HDL-C	rs328	8	2	rs10503669 ($r^2 = 1.0$) rs7007797 ($r^2 = 1.0$)	rs10503669 – $9.1*10^{-4}$ rs7007797 – $5.0*10^{-4}$
LPL	TG	rs328	8	2	rs10503669 ($r^2 = 1.0$) rs7007797 ($r^2 = 1.0$)	rs10503669 – 0.02 rs7007797 – 0.002
APOA5	TG	rs662799 rs3135506	4	0	-	-

A slew of genes linked to diseases were unearthed by Schadt and team (2008) by profiling a tissue-specific human liver cohort (HLC) aimed at comprehending the "genetic architecture of the liver". Analyses of 39,000 transcripts with 782,476 unique single nucleotide polymorphisms (SNPs) genotyped, DNA variants strongly associated with liver gene expression were discerned. rs599839 on Chromosome 1p13.3 identified as associated with coronary artery disease (CAD) in the Wellcome Trust Case Control Consortium (WTCCC) documented an association in the HLC of this work with *PSRC1*, *CELSR2*, and *SORT1* . Strong correlations emerged for *Psrc1*, *Sort1*, and *Celsr2* with plasma LDL-cholesterol levels in the BXH/wt cross (a cross aimed at specifically probing factors that elevate the risk of cardiovascular disease). Further, the rs599839 SNP was also correlated to higher LDL cholesterol levels. While lowered *CELSR2* and *SORT1* expression documented an association with elevated LDL levels in mice, *PSRC1* displayed the opposite trend of probably exerting a protective effect against CVD.

Figure 1 below documents SNP rs599839 implicated as a risk allele for CAD in the WTCCC (lilac panel). The pink panel illustrates the strong association of *Psrc1*, *Sort1*, and *Celsr2* with rs599839 and the green panel illustrates the cross to probe the metabolic risk of heart disease (© 2008 Schadt et al. *This is an open-access article distributed under the terms of the Creative Commons Attribution License, which permits unrestricted use, distribution, and reproduction in any medium, provided the original author and source are credited*):

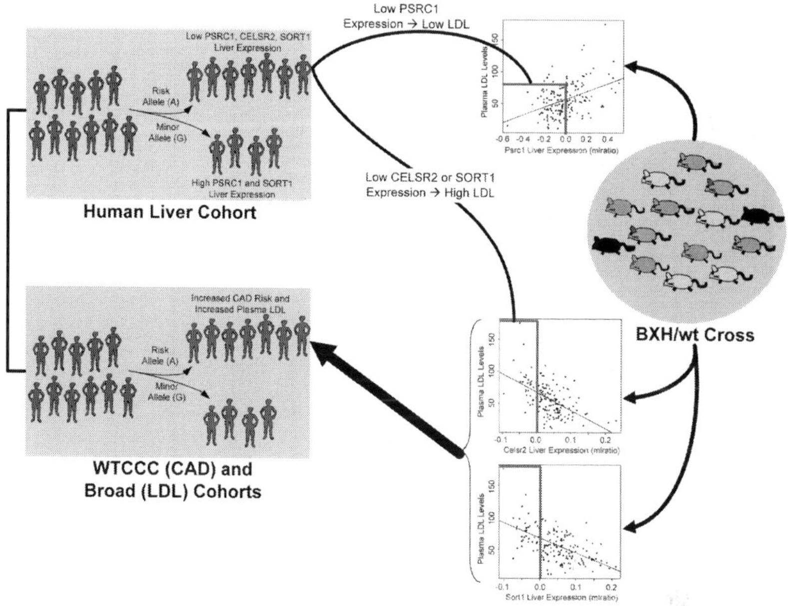

Chasman and team (2009) further corroborated the genetic influence of LDL, HDL and VLDL particle levels and distribution employing GWAS of 17 lipoprotein measures as outlined by NMR with LDL-C, HDL-C, triglycerides, ApoA1, and ApoB in 17,296 women from the Women's Genome Health Study (WGHS). Novel loci were reported: (*PCCB/STAG1* (3q22.3), *GMPR/MYLIP* (6p22.3), *BTNL2* (6p21.32), *KLF14* (7q32.2), 8p23.1, *JMJD1C* (10q21.3), *SBF2* (11p15.4), 12q23.2, *CCDC92/DNAH10/ZNF664* (12q24.31.B), and *WIPI1* (17q24.2)). Further, *LIPC* (15q22.1) and *APOC-APOE* complex (19q13.32) emerged as genetic determinants of total IDL and total VLDL, respectively. The loci unearthed by the team and their association with LDL-C, HDL-C, and triglycerides are depicted in table 3 below (© 2009 Chasman et al. *This is an open-access article distributed under the terms of the Creative Commons Attribution License, which permits unrestricted use, distribution, and reproduction in any medium, provided the original author and source are credited*):

						WGHS					Meta-analysis[a]		
Locus	gene(s)	sample	lead SNP[b]	lead fraction(s)[b]	internal replicationp-value[c]	Extende sample p-value[c]	LDL-C[d]	HDL-C[d]	TG[d]	N	LDL-C[d]	HDL-C[d]	TG[d]
3q22.3	*PCCB STAG1*	all	rs3856637	HDL:S (+)	0.018	**1.1×10^{-9}**	n.s.	1.1×10^{-2} (+)	n.s.	18516	0.14 (+)	0.20 (−)	3.0×10^{-5} (+)
6p21.32	*BTNL2*	all	rs2076530	VLDL:L (+)	0.34	1.6×10^{-7}	n.s.	n.s.	5.4×10^{-7} (+)	19584	0.94 (−)	0.034 (−)	0.33 (−)
				TG:N (+)	0.065	**1.3×10^{-9}**							
		fasting	rs3129882	VLDL:L (−)	0.057	**1.8×10^{-8}**	n.s.	5.9×10^{-4} (+)	2.1×10^{-7} (−)	19648	0.47(−)	7.0×10^{-3} (+)	0.35 (−)
				TG:N (−)	0.14	1.8×10^{-8}							
7q32.2	*KLF14*	all	rs4731702	LDL:S (−)	0.00019	5.3×10^{-9}	4.1×10^{-3} (−)	4.9×10^{-7} (+)	2.2×10^{-9} (−)	19648	0.56 (−)	8.8×10^{-7} (+)	5.9×10^{-3} (−)
				LDL:T (−)	0.0028	**5.4×10^{-11}**							
				HDL:Z (+)	0.0065	**1.0×10^{-10}**							
				TG (−)	0.0026	**2.7×10^{-11}**							
8p23.1	*Intergenic PPP1R3B*	fasting	rs983309	VLDL:Z (+)	0.14	**3.8×10^{-9}**	3.7×10^{-5} (−)	5.8×10^{-3} (−)	1.5×10^{-2} (+)	19648	2.4×10^{-4} (−)	5.7×10^{-6} (−)	0.52 (+)
12q23.2	*Intergenic ASCL1 PAH*	all	rs1818702	HDL:T (−)	0.04	**2.7×10^{-10}**	n.s.	n.s.	3.4×10^{-2} (−)	19536	0.88 (+)	0.99 (−)	0.069 (−)
		all	rs10778213[e]	HDL:N (−)	0.096	**7.6×10^{-9}**	0.021(+)	n.s.	n.s.	19648	0.12 (−)	0.22 (+)	0.47 (+)
12q24.31.B	*CCDC9 DNAH10 ZNF664*	all	rs7307277	HDL:L (+)	0.0032	**9.0×10^{-13}**	1.4×10^{-3} (−)	5.5×10^{-8} (+)	$6.1\times^{-10}$ (−)	19648	0.40(+)	0.074 (−)	0.35 (+)
				HDL:Z (+)	0.0013	**1.5×10^{-12}**							
				LDL:Z (+)	0.0073	**5.0×10^{-10}**							
				LDL:T (−)	0.00024	**1.1×10^{-11}**							
				LDL:S (−)	0.00029	**1.2×10^{-12}**							
				TG (−)	0.015	**6.7×10^{-11}**							
		fasting		HDL:L (+)	0.010	**1.4×10^{-9}**							
17q24.2	*PRKAR1A WIPI1*	all	rs2909207	HDL:M (+)	0.045	**1.7×10^{-9}**	n.s.	n.s.	2.2×10^{-3} (+)	19648	0.87 (−)	0.73 (+)	0.17 (+)

[a]Meta-analysis of LDL-C, HDL-C, and TG among 12 European populations [10].
[b]Most significant genome-wide SNP and lipoprotein fraction(s) association at locus for tests among the 22 lipoprotein fractions. (+/−) indicates trend of fraction with increasing copies of the minor allele (same as sign of beta coefficient, see Tables 2 and S2).
[c]P-value (two-sided) for association in additional 4639 (all) or 3305 (fasting) samples from the WGHS (internal replication). All association trends were consistent with discovery sample. Combining these new samples with the original discovery samples leads to p-values for the extended WGHS sample. Bold font indicates p-value smaller than in main discovery sample.
[d]P-value and trend (+/−) for association of increasing copies of the minor allele with indicated lipoprotein fraction in either WGHS or the meta-analysis. n.s. in WGHS analysis indicates not significant ($P>0.05$).
[e]Genome-wide significant association with plasma C-reactive protein in the WGHS [21].
doi:10.1371/journal.pgen.1000730.t003

Researchers Ma and team (2010) probed the effects of additive or dominance single-locus effects and epistasis on 432,096 SNP markers of total cholesterol (TC) and high-density lipoprotein cholesterol (HDL-C) in more than 6000 individuals of the Framingham Heart Study (FHS) data.

*Note: The FHS (launched in 1948) was originally aimed at probing common factors involved in cardiovascular disease.

Interestingly, new gene sets were unveiled: six new SNPs for the *RAB3GAP1-R3HDM1-LCT-MCM6* region of chr02 for TC and one novel SNP for the *OSBPL8-ZDHHC17* region (chr12) for HDL-C. Significant epistasis effects were documented for TC and HDL-C (two and thirteen, respectively) of which *LMBRD1* (chr06) and the *LRIG3* region (chr12) documented the most evident result for HDL-C (dominance × dominance effect: $p = 5.72 \times 10^{-13}$). A x A: additive × additive epistasis effect was noted for TC with chr04 and chr10 with the maximum TC value in the A-T gamete and the lowest TC value in the A-G gamete. In terms of the A × D: additive × dominance epistasis effect, the maximum TC value was in the A-GG allele-genotype and the lowest TC value was in the G-GG allele-genotype.

Figure 2 below illustrates the regions linked with cholesterol highlighted in quantitative genetics as reported by the researchers **A**) A 1.23 Mb region inclusive of *RAB3GAP1-R3HDM1-LCT-MCM6* showing multiple SNP effects on TC. **B**) A 1Mb region inclusive of *OSBPL8-ZDHHC17* linked to HDL-C. **C**) A 1 Mb region inclusive of *LMBRD1* with SNPs interacting with an SNP for TC near *LRIG3*. **D**) A 1 Mb region inclusive of *LRIG3* near an SNP for TC interacting with multiple SNPs in *LMBRD1* (Ma, Yang, Runesha

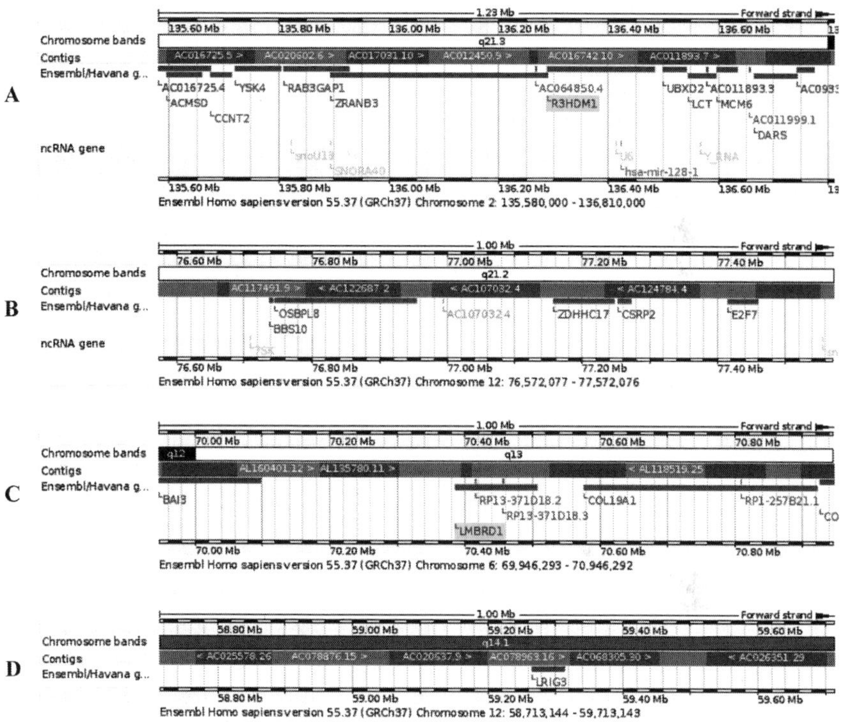

Researchers Wong and team (2020) probed the heritability of the plasma lipidome in 75 pairs of MZ twins and 55 pairs of DZ twins from the Older Australian Twin Study (OATS). The rationale for the classical twin design that approximates monozygotic and dizygotic twin pairs for additive genetic effects (A) and shared (C) and unique environmental (E) influences on a trait is that the matched age and environment in twins diminishes cross-generation differences. The lipid measures of total cholesterol, LDL, HDL and triglycerides documented significant heritability: h^2 = 0.427, 95% C.I. = [0.075, 0.592], 0.404, 95% C.I. = [0.121, 0.573] for total cholesterol, h^2 =0.404, 95% C.I. = [0.121, 0.573] for LDL, h^2 =0.419, 95% C.I. = [0.027, 0.766] for HDL and h^2 = 0.427, 95% C.I. = [0.181, 0.623] for triglycerides. A substantial shared/common environment aspect (C) emerged for HDL : h^2_C = 0.27, 95% C.I. = [0.00, 0.48].

The heritability estimates for the lipids probed by the researchers in twins as depicted by them is given in table 4 below (Matthew WK Wong, Anbupalam Thalamuthu, Nady Braidy et al, 2020 © 2020, Wong et al. *This article is distributed under the terms of the Creative Commons Attribution License, which permits unrestricted use and redistribution provided that the original author and source are credited*):

Study and cohort details	**Findings**	**Reference**
Traditional lipids		
Present study 75 MZ pairs, 55 DZ pairs 69–93 years	Range h^2: 0.404–0.427 HDL-C had substantial C-component 0.27	
Qingdao Twin Registry 382 MZ pairs and 139 DZ pairs, mean age 51 ± 7	Total Cholesterol and LDL-C 0.614, 0.655 HDL-C h^2 = 0.26, C-component = 0.478	Liu et al., 2018
National Heart Lung and Blood Institute Veteran Twin Study; 235 MZ, 260 DZ pairs 48–63 years	Longitudinal increases in heritability across three time pts Total Cholesterol (from 0.46 to 0.57), LDL-C (from 0.49 to 0.64), and HDL-C (from 0.50 to 0.62) TG: h^2 = 0.40	Goode et al., 2007
San Antonio Family Heart Study N = 569, mean age 39.4 years	h^2HDL-C = 0.55, h^2TG = 0.53	Goode et al., 2007; Mahaney et al., 1995
Lipid Species/Classes		
Present study	Range h^2: 0–0.59 Range heritable lipids: 0.287–0.59, median: 0.433 Heritable lipids: some Cer, TG, DG. Fewer PE, PC Non-heritable lipids: LPC, SM, PI, CE	
Wisconsin Registry for Alzheimer's Prevention n = 1212, mean age 60.8	Range 0.2–84.9%, median h^2 = 0.354 Median h^2ceramides = 0.48 Median h^2DG = 0.38	Darst et al., 2019
San Antonio Family Heart study, n = 1212 mean age 39.52	h^2range = 0.09–0.60 Median = 0.37 Heritable: almost all lipids, including Cer, TG, DG Least heritable: LPC, alkyl-PE	Bellis et al., 2014
NUGAT Twin Study 34 MZ, 12 DZ twin pairs 18–70 years, median age 25	Range h^2: 0–0.62 (19/152 lipids had h^2 > 0.40) Heritable lipids: LPC, PE, SM Non-heritable: Cer	Frahnow et al., 2017; Tabassum et al., 2019

FINRISK n = 2181 25–74 years SNP based heritability	SNP based range h^2: 0.10–0.54 Heritable lipids: Cer, LPC, SM, TG Non-heritable: PI	Tabassum et al., 2019
n = 203 plasma samples from 31 families	Cer heritability range: 0.10–0.63	McGurk et al., 2017
n = 999, 196 British families, mean age 45	SNP-based Cer heritability range: 0.18–0.87	McGurk et al., 2019

Another study probing the genetic aspects entailed twins but the sample set was inclusive of 82 twin pairs for heritability analysis and 139 dizygotic (DZ) twin pairs for GWAS from Qingdao Twin Registry (China). Analyses revealed that HDL-C variance could be explained by the ACE model with A at 26.6% (95% CI: 7.9–48.9), C at 47.8% (95% CI: 26.3–64.6) and E at 25.6% (95% CI: 20.8–31.6). The variance in TC and LDL-C could be explained by the AE model with the values depicted in table 5 below ():

Variable	Model	A (95%CI)		C (95%CI)		E (95%CI)		−2LL	df	AIC	χ^2	*P*
TC	ACE	51.8	(23.6–67.9)	9.0	(0–33.7)	39.2	(32.0–47.9)	1996.9	757	482.9		
	CE	–	–	49.5	(41.5–56.8)	50.5	(43.2–58.5)	2010.3	758	494.3	13.4	2.49E−04
	AE*	**61.4**	**(53.2–68.3)**	**–**	**–**	**38.6**	**(31.7–46.8)**	**1997.4**	**758**	**481.4**	**0.4**	**5.13E−01**
HDL-C	**ACE***	**26.6**	**(7.9–48.9)**	**47.8**	**(26.3–64.6)**	**25.6**	**(20.8–31.6)**	**1763.6**	**757**	**249.6**		
	CE	–	–	69.4	(63.7–74.4)	30.6	(25.6–36.3)	1771.7	758	255.7	8.1	4.48E−03
	AE	74.9	(69.5–79.4)	–	–	25.1	(20.6–30.5)	1778.5	758	262.5	14.9	1.16E−04
LDL-C	ACE	61.5	(34.9–71.7)	3.8	(0–27.4)	34.7	(28.3–42.7)	2028.1	760	508.1		
	CE	–	–	50.9	(43.0–58.0)	49.1	(42.0–57.0)	2049.4	761	527.4	21.2	4.06E−06
	AE*	**65.5**	**(57.9–71.8)**	**–**	**–**	**34.5**	**(28.2–42.2)**	**2028.2**	**761**	**506.2**	**0.09**	**7.71E−01**

**, The best fitted model; −2LL, −2 Log Likelihood; df, degree of freedom; χ^2, difference of χ^2-value; P, χ^2 test in model fitting; A, additive genetic influence; AIC, Akaike's information criterion; C, common or shared environmental influence; E, unique or non-shared environmental influence; HDL-C, high density lipoprotein cholesterol; LDL-C, low density lipoprotein cholesterol; TC, total cholesterol. The content discussed in detail were in bold.*

Few SNPs reached the requisite level of significance with gene-based analyses outlining a weak association of thousands of genes with all three parameters. The common genes for TC, LDL-C and HDL-C like *KLKB1* and *FAF1*. The pathways identified in pathway enrichment are depicted in figure 3 below ():

Pathway	chisq-*P*	emp-*P*	-log(chisq*P*)	-log(emp*P*)
REACTOME_CELL_SURFACE_INTERACTIONS_AT_THE_VASCULAR_WALL	7.11E-04	2.94E-04	3.14818	3.53165
BIOCARTA_HER2_PATHWAY	4.39E-04	4.35E-04	3.35736	3.36151
KEGG_**CELL_ADHESION_MOLECULES_CAMS**	6.25E-04	4.44E-04	3.20421	3.35262
REACTOME_**IL_6_SIGNALING**	6.27E-04	4.66E-04	3.20281	3.33161
KEGG_**TIGHT_JUNCTION**	2.91E-03	6.20E-04	2.53649	3.20761
BIOCARTA_AHSP_PATHWAY	8.78E-04	8.80E-04	3.05652	3.05552
REACTOME_**BILE_SALT_AND_ORGANIC_ANION_SLC_TRANSPORTERS**	9.23E-04	9.30E-04	3.03463	3.03152
REACTOME_**TRANSPORT_OF_GLUCOSE_AND_OTHER_SUGARS_BILE _SALTS_AND_ORGANIC_ACIDS_METAL_IONS_AND_AMINE_COMPOUNDS**	9.23E-04	9.50E-04	3.03463	3.02228
REACTOME_**SIGNALING_BY_ILS**	1.35E-03	1.17E-03	2.87021	2.93181
BIOCARTA_CHREBP2_PATHWAY	1.13E-03	1.30E-03	2.94743	2.88606
REACTOME_INTRINSIC_PATHWAY	1.51E-03	1.38E-03	2.82086	2.86012
REACTOME_FORMATION_OF_FIBRIN_CLOT_CLOTTING_CASCADE	1.51E-03	1.48E-03	2.82086	2.82974
REACTOME_NEPHRIN_INTERACTIONS	4.77E-03	1.82E-03	2.32139	2.73993
BIOCARTA_**IL6_PATHWAY**	2.04E-03	1.96E-03	2.69087	2.70774
KEGG_**NATURAL_KILLER_CELL_MEDIATED_CYTOTOXICITY**	7.95E-03	2.15E-03	2.09938	2.66756
REACTOME_**CHONDROITIN_SULFATE_BIOSYNTHESIS**	2.36E-03	2.31E-03	2.62634	2.63639
KEGG_**SULFUR_METABOLISM**	2.36E-03	2.34E-03	2.62634	2.63078
KEGG_**GLYCOSAMINOGLYCAN_BIOSYNTHESIS_CHONDROITIN_SULFATE**	2.36E-03	2.39E-03	2.62634	2.6216
REACTOME_APC_C_CDH1_MEDIATED_DEGRADATION_OF_CDC20_AND_OTHER_ APC_C_CDH1_TARGETED_PROTEINS_IN_LATE_MITOSIS_EARLY_G1	5.66E-03	2.61E-03	2.24755	2.58336
KEGG_**LEUKOCYTE_TRANSENDOTHELIAL_MIGRATION**	2.96E-03	2.62E-03	2.52912	2.5817

chisq-P, Chi-square p-value. Chi-squared method (gene-score p-value were ranked and transformed to a uniform distribution, these values were then transformed by a chi-square quantile function, and summed). The content discussed in detail were in bold.
emp-P, empirical p-value. Empirical sampling method (gene-scores are transformed with chi-square quantile function and summed, then Monte Carlo estimate of the p-values were obtained by sampling random sets of the same size).

Zimoń and team (2021) employed genetics and functional genomics i.e. protein-truncating variants (PTVs) from exome sequencing of 302,331 participants in the UK Biobank; ubiquitously employed SNPs genotyped from GWAS in the UK Biobank and RNAi-based combinatorial knockdown or combinatorial RNA interference (coRNAi) to score the uptake of LDL-C into cultured cells to score thirty genes involved in blood lipid level control. This was aimed at probing genetic interactions (GIs) and additive effects (AEs) where genes act additively within or between loci. The loss-of-function protein-truncating variants (PTVs)-based analyses led to the identification of *PCSK9-APOB* for LDLc and TC; *LPL-APOB* for HDLc, LDLc and TG; *LDLR-LPL* for LDLc; and *PCSK9-LPL* for LDLc as AEs. Further, the additive changes in plasma lipid levels emerged for carriers of PTVs in *APOB* and either *PCSK9* or *LPL*.

Figure 4 below depicts **a** The study plan with 30 candidate genes from 18 GWAS **b** PTV-based AEs between *PSCK9* and *APOB* for LDLc in n=215,810 **c** PTV-based AEs between *LPL* and *APOB* for HDLc in n=198,052 **d** between PTV-based AEs *LDLR* and *LPL* for LDLc in n=215,810, and **e** PTV-based AEs *PCSK9* and *LPL* for LDLc in n=215,810 where n= number of carriers. and (−)=predicted loss-of-function due to PTVs (Zimoń, Huang, Trasta et al, 2021: *Open Access: This article is licensed under a Creative Commons Attribution 4.0 International License, which permits use, sharing, adaptation, distribution and reproduction in any medium or format, as long as you give appropriate credit to the original author(s) and the source, provide a link to the Creative Commons license, and indicate if*

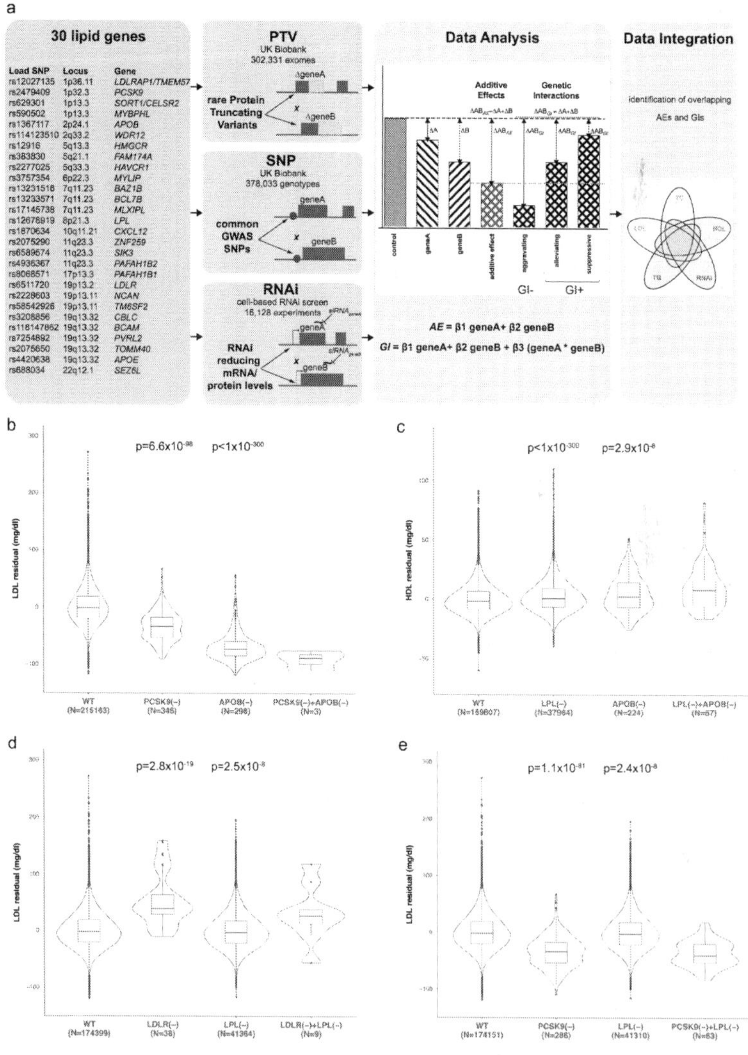
a
30 lipid genes
PTV
UK Biobank
302,331 exomes
rare Protein Truncating Variants
SNP
UK Biobank
378,033 genotypes
common GWAS SNPs
RNAi
cell-based RNAi screen
16,128 experiments
RNAi reducing mRNA/ protein levels
Data Analysis
Additive Effects
Genetic Interactions
control
geneA
geneB
additive effect
aggravating
alleviating
suppressive
GI-
GI+
AE = β1 geneA+ β2 geneB
GI = β1 geneA+ β2 geneB + β3 (geneA * geneB)
Data Integration
identification of overlapping AEs and GIs
b
p=6.6x10-98
p<1x10-300
LDL residual (mg/dl)
WT (N=215163)
PCSK9(-) (N=346)
APOB(-) (N=298)
PCSK9(-)+APOB(-) (N=3)
c
p<1x10-300
p=2.9x10-8
HDL residual (mg/dl)
WT (N=159807)
LPL(-) (N=37964)
APOB(-) (N=224)
LPL(-)+APOB(-) (N=57)
d
p=2.8x10-19
p=2.5x10-8
LDL residual (mg/dl)
WT (N=174399)
LDLR(-) (N=38)
LPL(-) (N=41364)
LDLR(-)+LPL(-) (N=9)
e
p=1.1x10-81
p=2.4x10-8
LDL residual (mg/dl)
WT (N=174151)
PCSK9(-) (N=286)
LPL(-) (N=41310)
PCSK9(-)+LPL(-) (N=63)

The RNAi-based assays unveiled interesting findings. Cellular LDLc uptake was augmented by simultaneous knockdown of *HMGCR* and *APOB* (aggravating GI) that was also seen with *PCSK9* and *TMEM57*, *MYBPHL* and *SIK3* as well as *SIK3* and *PAFAH1B1* (individual silencing exerted moderate effects). Simultaneous knockdown of *LDLR* and *LDLRAP1* caused a lower-than-expected reduction (positive alleviating GI).

Figure 5 below illustrates ***TOP***: **a** The coRNAi screening **b** Heatmap of gene-pair assessment: Red-increase and Blue-decrease with individual silencing in the top row and first column **c** positive alleviating and suppressive or negative aggravating GIs in 20 gene pairs **d–f** Single gene ($siRNA_{geneA}$+CTRL) and gene pair ($siRNA_{geneA}$+$siRNA_{geneB}$) siRNA knockdown effects on relative uptake of fluorescently-labeled LDLc (DiI-LDL) cellular uptake: CTRL, negative control siRNA (Zimoń, Huang, Trasta et al, 2021: *Open Access: This article is licensed under a Creative Commons Attribution 4.0 International License, which permits use, sharing, adaptation, distribution and reproduction in any medium or format, as long as you give appropriate credit to the original author(s) and the source, provide a link to the Creative Commons license, and indicate if changes were made. The images or other third party material in this article are included in the article's Creative Commons license, unless indicated otherwise in a credit line to the material. If material is not included in the article's Creative Commons license and your intended use is not permitted by statutory regulation or exceeds the permitted use, you will need to obtain permission directly from the copyright holder. To view a copy of this license, visit http://creativecommons.org/licenses/by/4.0/*):

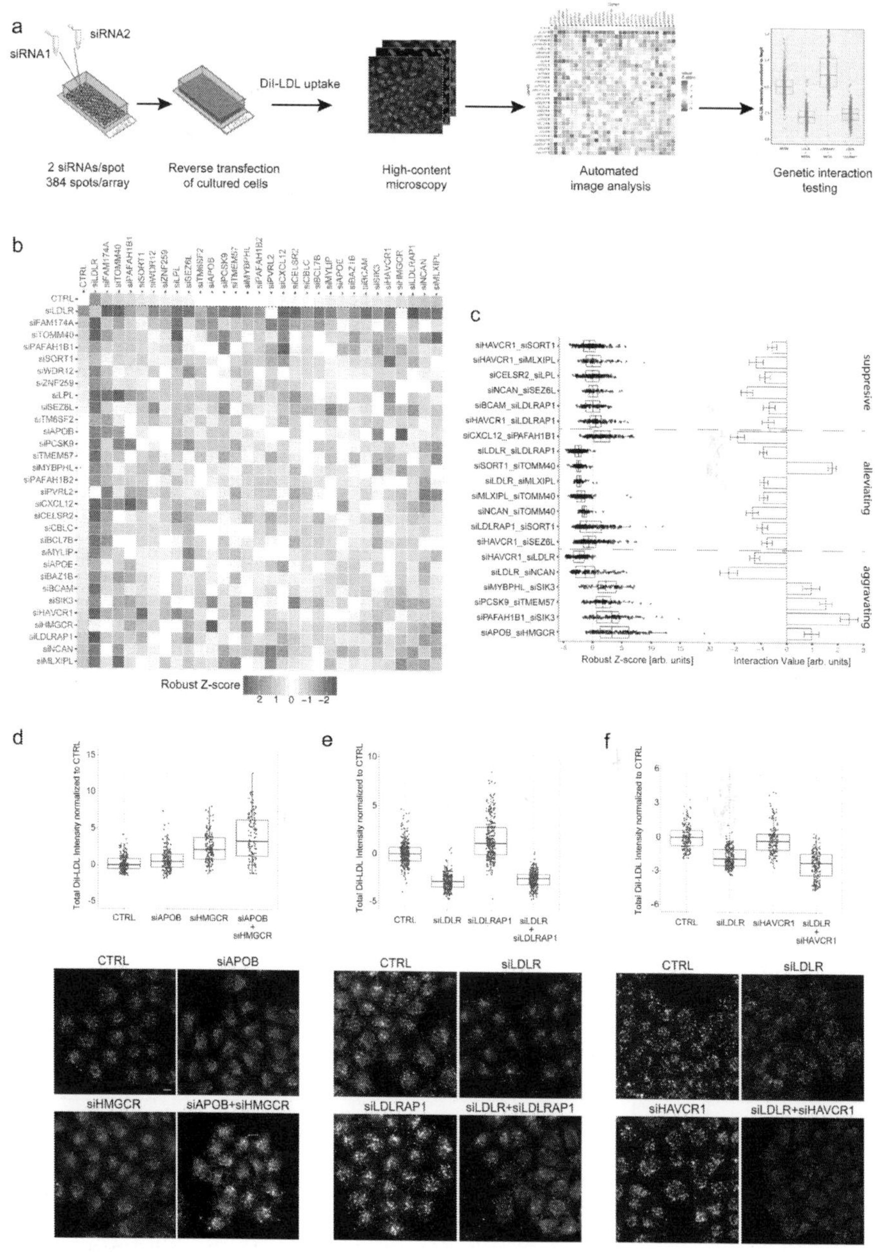

BOTTOM: Pair-wise gene interactions (GIs) for cellular LDL uptake (Zimoń, Huang, Trasta et al, 2021: *Open Access: This article is licensed under a Creative Commons Attribution 4.0 International License, which permits use, sharing, adaptation, distribution and reproduction in any medium or format, as long as you give appropriate credit to the original author(s) and*

GI Gene Pair		Interaction type		Primary RNAi screen			Validation RNAi screen		
Gene1	Gene2			Robust Z-score	Interaction Value	*p*-val(FDR)	Robust Z-score	Interaction Value	*p*-val(FDR)
APOB	*HMGCR*	NEGATIVE	Aggravating	2.8	2.18	1.47E−03	3.33	0.97	1.27E−03
HAVCR1	*LDLR*			−2.18	−1.32	5.99E−03	−2.24	−1.24	2.72E−10
LDLR	*NCAN*			−1.86	−1.28	3.56E−03	−1.23	−2.22	3.39E−11
MYBPHL	*SIK3*			2.17	1.59	8.55E−04	2.3	0.96	5.85E−03
PAFAH1B1	*SIK3*			1.62	1.79	4.50E−04	3.3	2.44	7.06E−12
PCSK9	*TMEM57*			1.46	1.76	5.27E−04	1.79	1.53	1.89E−09
HAVCR1	*SEZ6L*	POSITIVE	Alleviating	−1.19	−1.91	8.19E−05	−0.67	−0.77	3.82E−04
LDLR	*LDLRAP1*			−2.44	−1.86	6.66E−06	−2.49	−0.92	5.18E−10
LDLR	*MLXIPL*			−2.03	−1.49	4.14E−04	−2.31	−0.90	3.79E−08
LDLRAP1	*SORT1*			−1.11	−1.65	7.94E−04	−1.09	−0.95	6.18E−06
MLXIPL	*TOMM40*			−2.14	−2.67	5.38E−08	−1.94	−0.90	3.79E−08
*NCAN**	*TOMM40**			−1.4	−1.56	3.23E−03	−1.49	−1.34	3.45E−08
*SORT1**	*TOMM40**			0.77	1.84	9.95E−04	−2.48	1.77	<1.00E−15
BCAM	*LDLRAP1*		Suppressive	−0.4	−1.83	3.31E−05	−0.05	−0.7	4.03E−03
CELSR2	*LPL*			−1.4	−1.43	3.24E−03	−0.11	−0.86	8.11E−05
CXCL12	*PAFAH1B1*			−2.17	−1.78	1.13E−04	1.32	−1.92	1.77E−12
HAVCR1	*LDLRAP1*			−0.56	−2.16	6.87E−07	0.39	−0.73	4.12E−03
HAVCR1	*MLXIPL*			−0.63	−2.31	1.38E−07	−0.12	−1.2	2.07E−05
HAVCR1	*SORT1*			−2.09	−2.24	2.26E−05	−0.63	−0.58	2.39E−03
NCAN	*SEZ6L*			−0.59	−1.6	1.18E−03	−0.08	−1.56	1.26E−11

Such studies are opening up the avenue of looking at pair-wise control of lipid genes for controlling blood lipid levels in the clinic.

4.2 GENETIC CAUSES

Familial hypercholesterolemia (FH) is a ubiquitous autosomal inherited disorder documenting elevated low-density lipoprotein cholesterol (LDL-C) profiles (linked to coronary artery disease or CAD) and affects one in 250 individuals. Genetically, the condition is attributed to the LDL receptor (*LDLR*) mutations in 90% of the cases along with reports of mutations in apo-lipoprotein B (*APOB*) in 5–10% cases and 1% of the cases have documented mutated protein convertase subtilisin/kexin type 9 (*PCSK9*) mutations. These variants cause diminishing of LDL protein clearance from the blood that causes an elevation in LDL-C levels in FH patients: reaching 9–12 mmol/l in heterozygous patients. Such augmented levels of LDL-C have been associated with atherosclerosis as per research. The LDL-C levels in FH patients are demonstrative of variability given that 1700 and more structural variants have been documented in the LDLR that impact the structure and functioning of the receptor differently.

(Truong, Do, Kim, Nguyen, Le and Le, 2020; Vrablik, Tichý, Freiberger, Blaha, Satny and Hubacek, 2020)

Table 6 below documents FH and its genetics (© 2020 Vrablik, Tichý, Freiberger, Blaha, Satny and Hubacek. *This is an open-access article distributed under the terms of the Creative Commons Attribution License (CC BY). The use, distribution or reproduction in other forums is permitted, provided the original author(s) and the copyright owner(s) are credited and that the original publication in this journal is cited, in accordance with accepted academic practice. No use, distribution or reproduction is permitted which does not comply with these terms*):

Gene	Protein	Protein function	Mode of inheritance	Resulting effect
LDLR	Low-density lipoprotein receptor	Cell surface receptor that plays an important role in cholesterol homeostasis, transporting cholesterol through cell surface	AD	↓ binding affinity of LDL-receptor protein
APOB	Apolipoprotein B	Major protein of LDL particle, ligand for LDL receptor	AD	↓ binding affinity of APOB to LDL-receptor
PCSK9	Proprotein-convertase subtilisin kexin 9	Protease that plays a role in LDL-receptor degradation	AD	↑ LDL-receptor protein degradation
STAP1	Signal-transducing adaptor protein-1	Unknown	AD	Unknown/incomplete association with hypercholesterolemia
LDLRAP1	Low-density lipoprotein receptor adaptor protein 1	Adaptor protein that interacts with cytoplasmatic tail of LDL receptor, promoting LDL particle internalization	AR	LDL-receptor dysfunction
LIPA	Lysosomal acid lipase alias cholesterol esters lipase	Hydrolysis of cholesterol esters or triglycerides	AR	APOB overproduction, upregulation of HMGCoA reductase
ABCG5/ABCG8	Sterolin-1 and 2	Transporters required for secretion of cholesterol into bile	AR	↑ absorption of plant sterols
PNPLA5	Patatin-like phospholipase domain-containing protein 5	Influencing adipocyte differentiation, triglyceride hydrolysis	AR	Possibly lipolysis impairment

AR, autosomal recessive; AD, autosomal dominant; HMGCoA, hydroxy-methyl glutaryl coenzyme A.

with accepted academic practice. No use, distribution or reproduction is permitted which does not comply with these terms):

Researchers	*PCSK9* variants	Population	Sample size	Variant frequency, %	LDL-C ranges, mmol/L
Cohen et al. (2006)	Y142X	ARIC study, general	3,363	0.8	1.7–3.7
	C679X	ARIC study, general	3,363	1.8	1.4–3.8
	R46L	ARIC study, general	9,524	3.2	2.2–3.9
Hooper et al. (2007)	C679X	African, general	653	3.7	1.3–1.9
Scartezini et al. (2007)	R46L	British, general	2,444	1.0*	1.8–4.0
Guella et al. (2010)	R46L	Italian, MI patients	1,880	1.0*	2.1–3.9
Chernogubova et al. (2012)	R46L	Swedish, general	5,722	1.9*	2.5–4.9
Saavedra et al. (2014)	R46L	Canadian, FH	562	3.0	5.8–7.7
Langsted et al. (2016)	R46L	CGPS study	103,083	1.3*	2.2–3.4
Mostaza et al. (2018)	R46L	Spanish, adults	1,188	2.9*	2.6–4.3
	R46L	Spanish, children and adolescents	1,933	3.2*	2.2–2.6

*Normal range of LDL-C: Low-risk groups < 3 mmol/L; High-risk groups < 1.8 mmol/L; FH groups < 1.4 mmol/L. (According to the 2019 ESC/EAS guidelines for the management of dyslipidaemias). PCSK9, proprotein convertase subtilisin/kexin type 9; LDL-C, low-density lipoprotein cholesterol; ARIC, Atherosclerosis Risk in Communities; MI, myocardial infarction; CGPS, Copenhagen General Population. *Represents the allele frequency which is different from the PCSK9 variant proportion of the population in the different references.*

The varying impact of LDLR protein variants specifically with reference to the six-bladed beta-propeller of epidermal growth factor precursor homology domain of the protein was probed by Dušková and team (2020) for structure, localization and impact on specific ER proteins. Analyses unveiled three LDLR protein variant groups:

1) p.Val429Met, p.Gly478Arg, p.Gly565Val, and p.Ser610Cys were mutations within the protein structure that boosted clumping of ER proteins and elevated ER-resident chaperones Hsp90b1 (heat-shock 90-kD protein beta 1), Hspa5 (heat-shock 70-kD protein 5), and Calr (calreticulin) while conspicuously lowering the internalization of LDL particles.
2) p.Glu408Val, p.Arg416Trp, and p.His583Arg emerged as surface mutations on the 3D structure that slightly impacted the LDLR localization in the plasma membrane.
3) p.Tyr532Cys, p.Leu555Pro, and p.His583Tyr were also surface mutations but they boosted ER resident chaperone mRNA levels on the similar lines of the first group of variants.

Such studies pave the way for further comprehension of the impact of varied mutations on the structure and functioning of the LDLR in FH.

Figure 6 below depicts the findings of the researchers who probed the impact of missense mutations reported in the LDLR receptor in FH. ***TOP***: Summary of the scrutiny

Protein variant	Contact via side chain	Buriedness (RSA)	Change of			Pro/Gly change	LCIM	FC (loc., % of wt)	FC (LDL int., % of wt)	Increased chaperone expres-sion	Group
			Charge	Polarity	Volume						
wt	NA	NA	NA	NA	NA	NA	Membrane + ER	100 ± 9	100 ± 17	No	–
p.Glu408Val		26%					Membrane + ER	74 ± 15	64 ± 11	No	2
p.Arg416Trp		54%					Membrane + ER	72 ± 1	78 ± 15	No	2
p.Val429Met		6%					ERa	0	0	Yes	1
p.Gly478Arg		0%					ERa	42 ± 4	64 ± 3	Yes	1
p.Tyr532Cys		14%					ERa	71 ± 9	67 ± 9	Yes	3
p.Leu555Pro		18%					ERa	52 ± 12	49 ± 12	Yes	3
p.Gly565Val		0%					ERa	0	0	Yes	1
p.His583Arg		19%					Membrane + ER	77 ± 3	80 ± 6	No	2
p.His583Tyr		19%					Membrane + ERa	40 ± 3	62 ± 12	Yes	3
p.Ser610Cys		0%					ERa	52 ± 6	56 ± 4	Yes	1

The blue fields show detected structure features (see section "Materials and Methods" for definition of features). ER: endoplasmic reticulum; ERa: accumulation of LDLR on ER in the form of clumps; LCIM: live cell imaging microscopy; FC (loc.): LDLR plasma membrane localization determined by flow cytometry; FC (LDL int.): LDL particle internalization determined by flow cytometry; wt: wild-type; Group: three groups described in the study (see section "Discussion" for definition of groups).

BOTTOM: Live cell imaging of **(A)** the LDLR (green: EYFP) and ER (red: EFP): Hoechst 33342 nuclear staining (blue) **(B)** the LDLR (green: EYFP) and LDL particles (red: pHRodo-Red) in the LDLR mutations in FH; Scale bar represents 10 μm

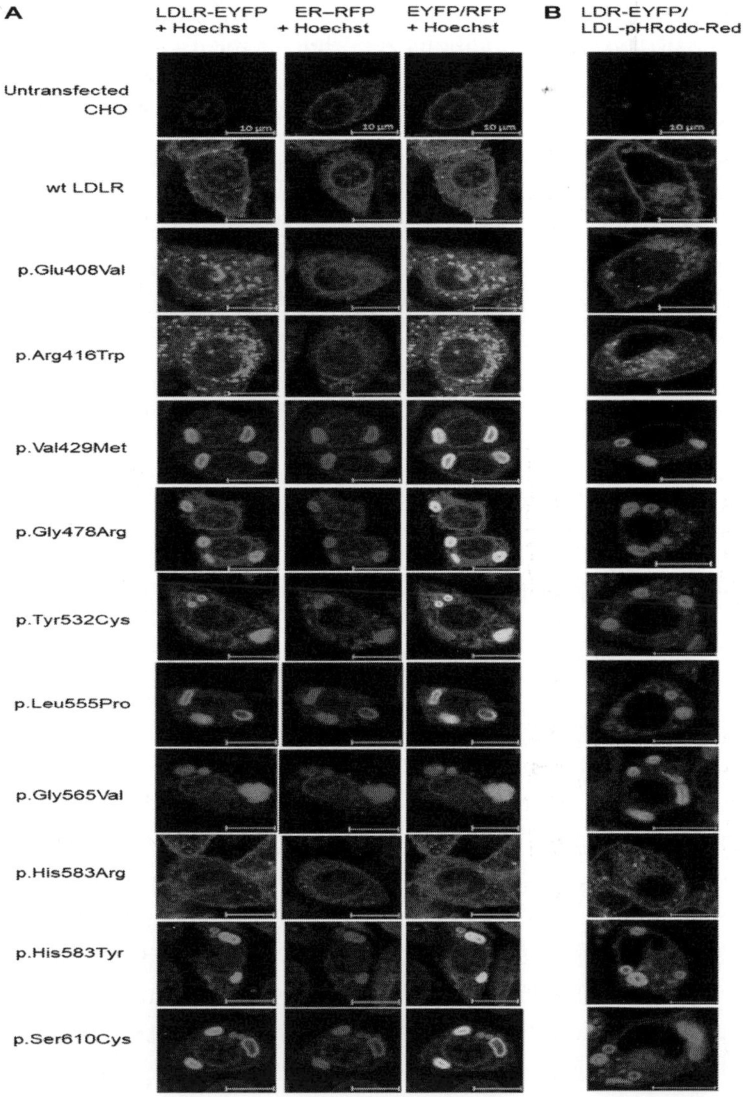

4.3 GENES AND STATIN RESPONSE

The management of cardiovascular disease entails the use of statins or 3-hydroxymethyl-3-methylglutaryl coenzyme A (HMG-CoA) reductase inhibitors. While this approach has achieved a lowering of LDL-C by 55% and cardiovascular events by 20-30%, the response to these drugs is varied (Postmus, Trompet, Deshmukh et al, 2014).

In order to probe the "genetic" angle behind this variation, genotyping of 148 single-nucleotide polymorphisms (SNPs) within 10 candidate genes related to lipid metabolism in patients treated with pravastatin (n=1536) was conducted by Chasman and team (2004). Analyses unveiled two SNPs in the gene for in the gene coding for 3-hydroxy-3-methylglutaryl-coenzyme A (HMG-CoA) reductase that is targeted by the drug. Both the variations lowered the efficacy of the statin: a diminishing of LDL-C reduction by 19% and a lowering of total cholesterol reduction due to the drug by 22% in individuals with a single copy of the minor allele vs. those who had the major alleles.

A 2014-published article by Postmus, Trompet, Deshmukh et al (2014) described as "the largest meta-analysis of GWAS for LDL-C response to statin therapy conducted to date" on the basis of a constructed Genomic Investigation of Statin Therapy (GIST) consortium with 6 RCTs (n=8,421 statin recipients) and 10 observational studies (n=10,175 statin recipients). Genome-wide significance ($P<5 \times 10^{-8}$) was demonstrated by the rs445925 SNP on *APOE*: minor allele frequency (MAF)=0.098 and $\beta=-0.043$ indicative of a lowering of 4.3% per allele in rs445925 SNP carriers for the LDL lowering due to statins. Another association was rs10455872 at *LPA*: MAF=0.069 and $\beta=0.041$ indicative of a lowering of 5.9 per allele in rs445925 SNP carriers for the LDL lowering due to statins. The third finding was the rs13166647 SNP at *RICTOR* on chromosome 5 (MAF=0.230, $\beta=-0.253$). SNPs at two new loci emerged with subsequent analyses: rs646776 on *SORT1/CELSR2/PSRC1* and rs2900478 on *SLCO1B1*. These were indicative of *SORT1/CELSR2/PSRC1* SNP carriers documenting a 1.5% increase per allele in LDL-C reduction and a decrease in LDL-C reduction by 1.6% per minor allele for *SLCO1B1* SNP carriers.

The results of the genetic insights accounting for varied responses to statins are documented in table 9 below and figure 7 below as reported by the scientists.

Table 9: GWAS analyses (Postmus, Trompet, Deshmukh et al, 2014: *This work is licensed under a Creative Commons Attribution 4.0 International License. The images or other third party material in this article are included in the article's Creative Commons license, unless indicated otherwise in the credit line; if the material is not included under the Creative Commons license, users will need to obtain permission from the license holder to reproduce the material. To view a copy of this license, visit http://creativecommons.org/licenses/by/4.0/*):

Key : Chr, chromosome; SNP, single nucleotide polymorphism; † % of extra LDL-C lowering in SNP carriers vs. non-carriers

Chr	Position	Lead SNP	Gene	Coding allele	Noncoding allele	Phase	*N*	Frequency-coding allele	Beta*	s.e.	% Extra reduction†	*P* value
1	109620053	rs646776	*SORT1/CELSR2/PSRC1*	C	T	Stage 1	16,697	0.230	−0.015	0.003	1.5	6.70×10^{-7}
						Stage 2	21,902	0.216	−0.010	0.003	1.0	2.43×10^{-4}
						Combined	38,599		−0.013	0.002	1.3	1.05×10^{-9}
6	160930108	rs10455872	*LPA*	G	A	Stage 1	12,981	0.069	0.041	0.006	−4.1	1.95×10^{-11}
						Stage 2	18,075	0.087	0.059	0.005	−5.9	7.14×10^{-35}
						Combined	31,056		0.052	0.004	−5.2	7.41×10^{-44}
12	21260064	rs2900478	*SLCO1B1*	A	T	Stage 1	16,749	0.165	0.016	0.003	−1.6	2.26×10^{-6}
						Stage 2	7,504	0.164	0.017	0.006	−1.7	3.54×10^{-3}
						Combined	24,253		0.016	0.003	−1.6	1.22×10^{-9}
19	50107480	rs445925	*APOE*	A	G	Stage 1	13,909	0.098	−0.043	0.005	4.3	1.58×10^{-18}
						Stage 2	3,613	0.157	−0.088	0.011	8.8	1.41×10^{-15}
						Combined	17,522		−0.051	0.005	5.1	8.52×10^{-29}

Figure 7: The loci in the combined meta-analysis (*n*=40,914) for (**a**) the *APOE* locus (**b**) the *LPA* locus (**c**) the *SORT1/CELSR2/PSRC1* locus and (**d**) the *SLCO1B1* locus: the top SNP is in purple Postmus, Trompet, Deshmukh et al, 2014: *This work is licensed under a Creative Commons Attribution 4.0 International License. The images or other third party material in this article are included in the article's Creative Commons license, unless indicated otherwise in the credit line; if the material is not included under the Creative Commons license, users will need to obtain permission from the license holder to reproduce the material. To view a copy of this license, visit http://creativecommons.org/licenses/by/4.0/*):

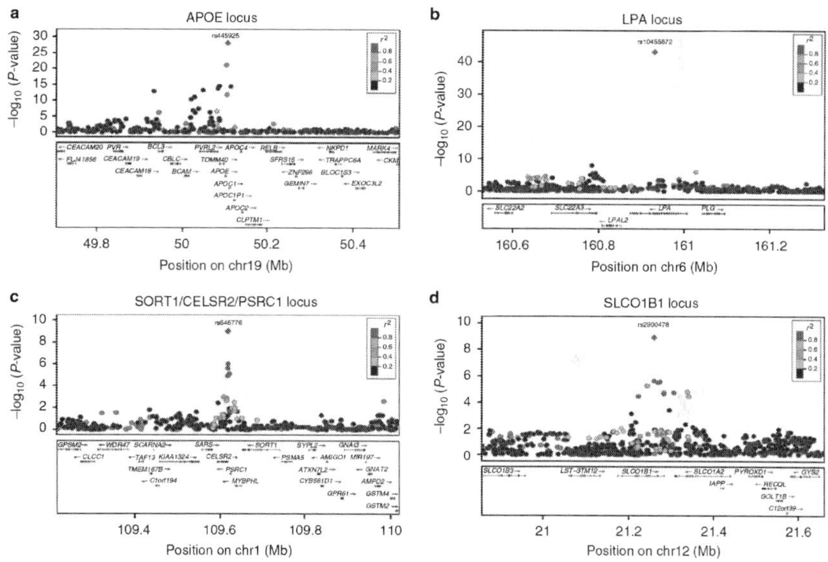

In a nutshell, such studies pave the way for further elucidation and comprehension of the genetics behind "personalized" treatment of statins to control LDL-C levels.

4.4 GENDER AND CHOLESTEROL PROFILES:

"It is time that we all see gender as a spectrum instead of two sets of opposing ideals."

— *Emma Watson*

A 2009-published cross-sectional epidemiological study with 209 participants (44 to 66 years old) in an urban Tanzanian population by Njelekela and team explored several factors of metabolism inclusive of cholesterol. The scrutiny revealed no evident difference for hypercholesterolemia (≥6.2

mmol/L) and higher LDL-cholesterol (≥3.8 mmol/L) between males and females but the odds of metabolic syndrome ($p = 0.001$) in terms of lowered HDL-C were higher in women over men by a factor of 3.

Figure 8 below illustrates the profile of CVD risk factors in males and females in the African population assessed: $P < 0.05$ and age-adjusted (Njelekela, Mpembeni and Muhihi et al, 2009: This article is published under license to BioMed Central Ltd. *This is an Open Access article distributed under the terms of the Creative Commons Attribution License (http://creativecommons.org/licenses/by/2.0), which permits unrestricted use, distribution, and reproduction in any medium, provided the original work is properly cited*):

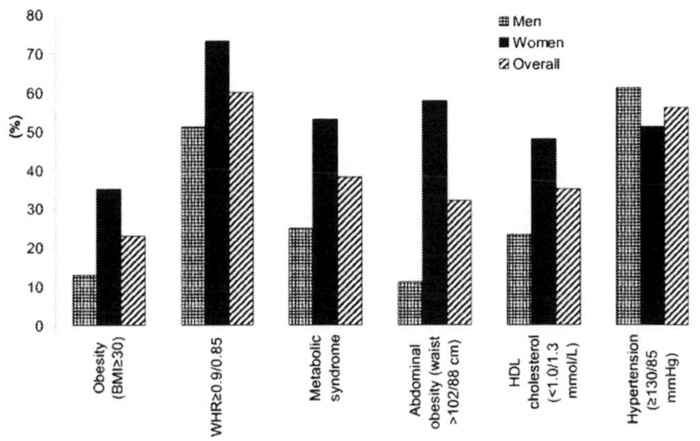

Research published in the International Journal of Endocrinology by Russo and team (2015) on LDL-C level management in type 2 diabetes (T2DM) patients probed 415.294 patients (45.3% women) from 236 diabetes centers in Italy. Analyses unveiled higher total cholesterol, LDL-C, and HDL-C serum levels in women as opposed to men with the lipid monitoring also lower in females. In those on lipid-lowering medications, the LDL-C target was not met by more females than males with age and duration of diabetes also boosting this gap. Figure 9 below illustrates the LDL-C management as documented by the researchers (Giuseppina Russo, Basilio Pintaudi, Carlo Giorda, Giuseppe Lucisano, Antonio Nicolucci, Maria Rosaria Cristofaro, Concetta Suraci, Maria Franca Mulas, Angela Napoli, Maria Chiara Rossi, Valeria Manicardi, 2015: © 2015 Giuseppina Russo et al. *This is an open access article distributed under the Creative Commons Attribution License, which permits unrestricted use, distribution, and reproduction in any medium, provided the original work is properly cited*):

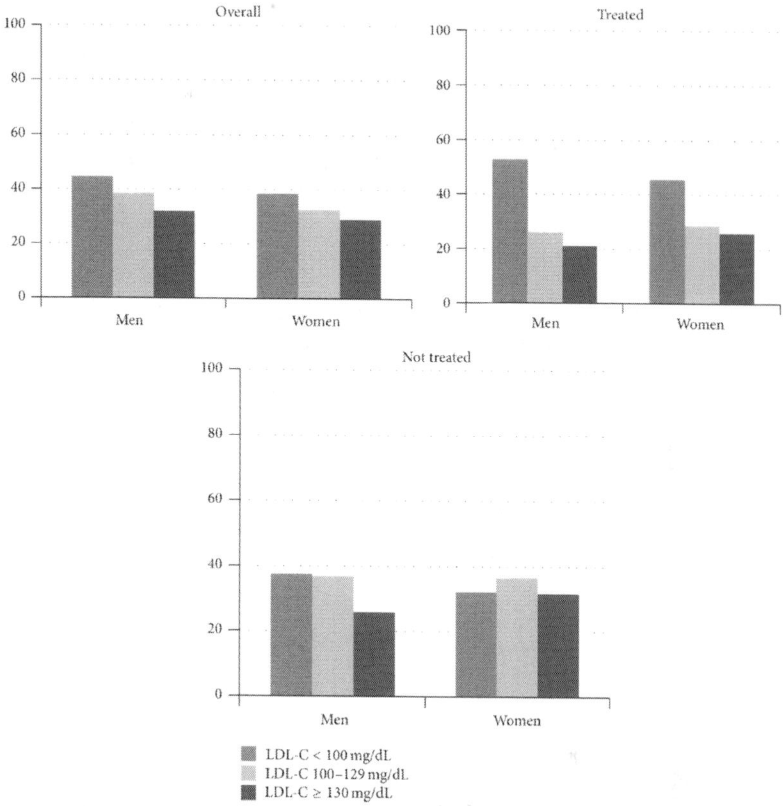

The management of optimal lipid profiles between the genders was probed by Zhang and team (2017) in 1721 male and 2072 female type 2 diabetes outpatients from the China Cardiometabolic Registries (CCMR)-3B study. As opposed to the goal of LDL-C <1.8 mmol/L, the levels were more in women as opposed to men (2.82 vs. 2.54 mmol/L; $p<0.001$). This trend was also seen for total cholesterol (4.98 vs. 4.46 mmol/L in women vs. men; $p<0.001$). Only 38.1% of women received lipid lowering therapy as opposed to 48.2% of the men ($p<0.001$). Such discrepancies in this population assayed as well as the abovementioned report warrant the addressing of gender imbalance in therapy and management too.

Figure 10 below illustrates the differences in lipid management between the genders in the study employing CCMR-3B data (Zhang, Ji and Ran et al, 2017: *Open Access: This article is licensed under a Creative Commons Attribution 4.0 International License, which permits use, sharing, adaptation, distribution and reproduction in any medium or format, as long as you give appropriate credit to the original author(s) and the source, provide a link to the Creative Commons license, and indicate if changes were made.*

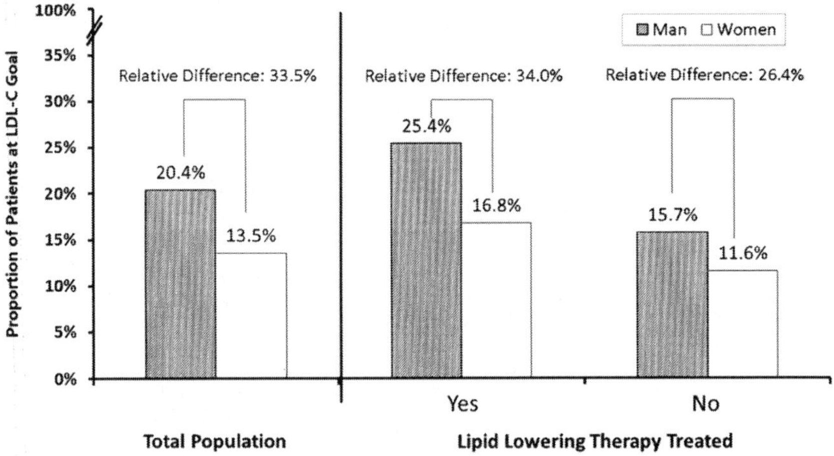

Another study in *Lipids in health and disease* by Wang and team (2020) explored the trends in dyslipidemia in rural Chinese individuals (n=26,378; 45–69 years old) from the Endoscopic Screening for Esophageal Cancer in China (ESECC) data. Analyses unveiled that women documented evidently lowered high-density lipoprotein cholesterol but elevated borderline high and above (BHA) LDL-C and total cholesterol as opposed to males. Interestingly, these levels increased with age in women. This dyslipidemia was linked to hypertension mainly in females and obesity in men. Figure 11 below depicts the trends of dyslipidemia with age and gender in the rural Chinese cohort analyzed between 2012 and 2016: a) borderline high and above (BHA) dyslipidemia b) high dyslipidemia c) Total cholesterol (TC) profiles d) Triglyceride (TG) profile 3)BHA low-density lipoprotein cholesterol (LDL-C) f) high-density lipoprotein cholesterol (HDL-C) (Wang, Liu, Li et al, 2020:

in this article are included in the article's Creative Commons licence, unless indicated otherwise in a credit line to the material. If material is not included in the article's Creative Commons licence and your intended use is not permitted by statutory regulation or exceeds the permitted use, you will need to obtain permission directly from the copyright holder. To view a copy of this licence, visit http://creativecommons.org/licenses/by/4.0/. The Creative Commons Public Domain Dedication waiver (http://creativecommons.org/publicdomain/zero/1.0/) applies to the data made available in this article, unless otherwise stated in a credit line to the data):

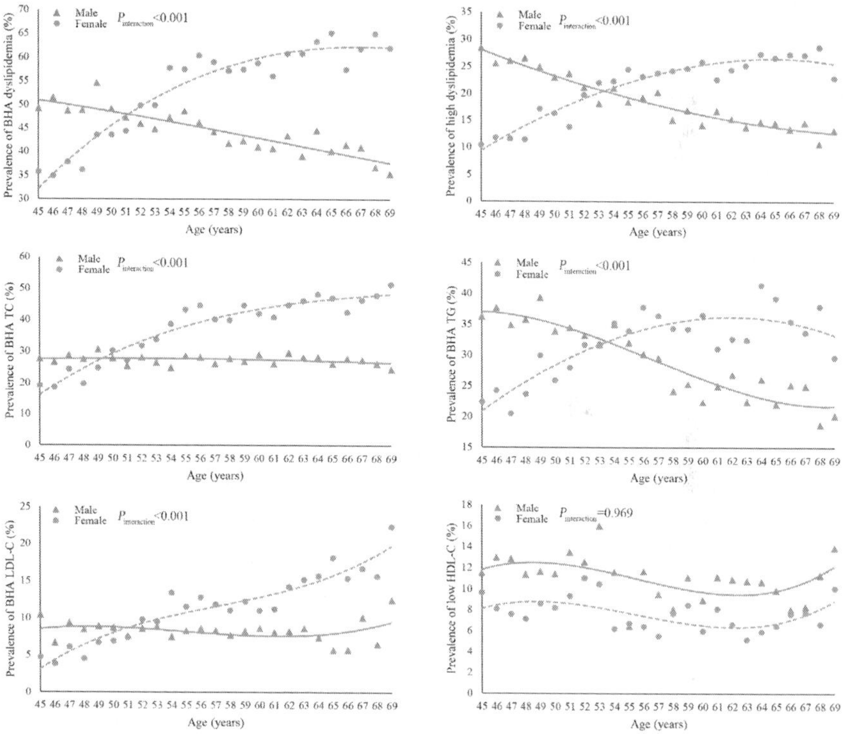

Tran et al (2021) documented an increase in total cholesterol (TC) in women with a rise at age 50 where the levels exceeded those in men in 25–64-year-old participants from Vietnam (n = 14706, 50.9% females). The spike in TC levels in Vietnamese women is documented by the researchers in figure 12 below (Tran, Blizzard and Luong et al, 2021: © 2021 Tran et al: *This is an open access article distributed under the terms of the Creative Commons Attribution License, which permits unrestricted use, distribution, and reproduction in any medium, provided the original author and source are credited*):

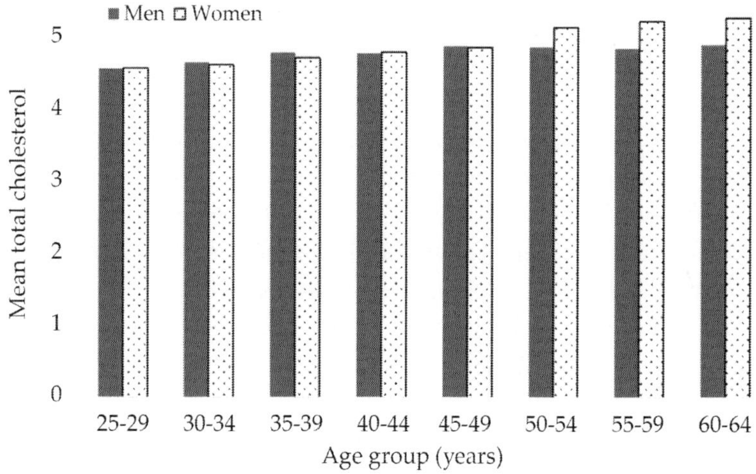

Karvinen and team (2019) scrutinized a cohort of the Estrogenic Regulation of Muscle Apoptosis (ERMA) study with 47–55-year-old healthy women from Finland for leisure time physical activity (LTPA) and the lipid profile. This age group is pertinent given the rise in cardiovascular disease (CVD) risk post-menopause. It emerged that the post-menopausal group documented elevated serum total cholesterol vs. that of the peri-menopausal and pre-menopausal groups while the peri-menopausal group documented higher total cholesterol vs. the pre-menopausal group. This trend was also documented for LDL-C and HDL-C.

Figure 13 below depicts the cholesterol trend in Total cholesterol (A), LDL-C (B), HDL-C (C), and triglyceride levels (D) in the pre-, peri- and post-menopausal groups: Recommended levels: Total cholesterol <5.2 mmol/L (<200 mg/dl), LDL <2.6 mmol/L (<100 mg/dl), HDL >1.5 mmol/L (60 mg/dl), and triglycerides <1.7 mmol/L (<150 mg/dl) (Mayo clinic, USA). Figures show average + SD, $*p < 0.050$, $***p < 0.001$ ():

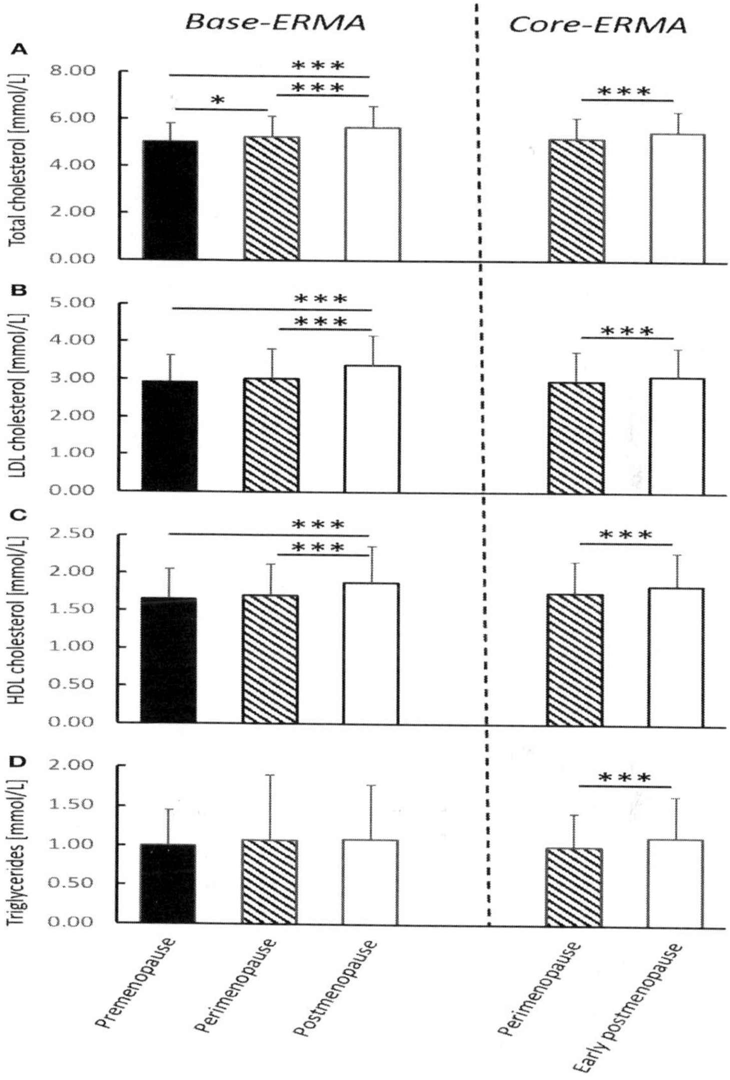

4.5 FINAL POINTS

The initial sections of this chapter elaborated research findings on how genes and gene interactions impact the cholesterol profile and its management. To summarize the role of gender, in the cohorts probed by various researchers across genders of different ethnicities, women seem to demonstrate a higher trend of cholesterol profiles. Few of the abovementioned studies also documented bad management of lipid profiles more in ladies vs. that in men.

Saikia et al (2016) scrutinized the healthcare expenditure (HCE) in India employing two rounds of nationally representative panel data—the India Human Development Survey (IHDS): I and II: 2004–2005 and 2011–2012. Both the prevalence of short-term and major morbidity was more in women while the mean health expenditure for women was always lower than men. Figure 14 below illustrates the mean HCE in both genders as analyzed by the authors (Saikia N, Moradhvaj, Bora JK, 2016: © 2016 Saikia et al. *This is an open access article distributed under the terms of the Creative Commons Attribution License, which permits unrestricted use, distribution, and reproduction in any medium, provided the original author and source are credited*):

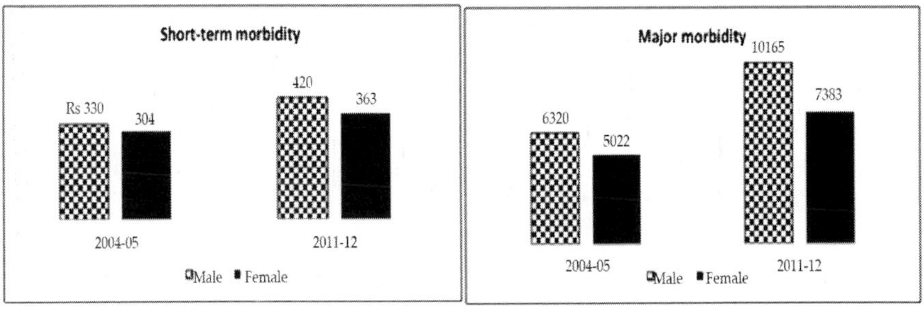

Thus, in addition to maintaining optimal cholesterol profiles, another avenue is opened here-the question of gender equality. Ongoing research can shed light on the efficient management of cholesterol across nationalities and genders.

"We can all agree that we want our institutions to grow and innovate and provide the best possible outcomes for our communities. The same can be said for our journals and our professional societies and meetings. The way to ensure peak performance across the board is to eliminate gender disparities. Diversity, equity, and inclusion in leadership and thought strengthen all of us."

- Spector and Overholser (2019)

4.6 REFERENCES

1. Kathiresan, S., Manning, A.K., Demissie, S. et al. A genome-wide association study for blood lipid phenotypes in the Framingham Heart Study. BMC Med Genet 8, S17 (2007). https://doi.org/10.1186/1471-2350-8-S1-S17

2. Bennet AM, Di Angelantonio E, Ye Z, et al. Association of Apolipoprotein E Genotypes With Lipid Levels and Coronary Risk. *JAMA*. 2007;298(11):1300–1311. doi:10.1001/jama.298.11.1300

3. Schadt EE, Molony C, Chudin E, Hao K, Yang X, Lum PY, et al. (2008) Mapping the Genetic Architecture of Gene Expression in Human Liver. PLoS Biol 6(5): e107. https://doi.org/10.1371/journal.pbio.0060107

4. Chasman DI, Paré G, Mora S, Hopewell JC, Peloso G, Clarke R, et al. (2009) Forty-Three Loci Associated with Plasma Lipoprotein Size, Concentration, and Cholesterol Content in Genome-Wide Analysis. PLoS Genet 5(11): e1000730. https://doi.org/10.1371/journal.pgen.1000730

5. Ma, L., Yang, J., Runesha, H.B. et al. Genome-wide association analysis of total cholesterol and high-density lipoprotein cholesterol levels using the Framingham Heart Study data. BMC Med Genet 11, 55 (2010). https://doi.org/10.1186/1471-2350-11-55

6. Matthew WK Wong, Anbupalam Thalamuthu, Nady Braidy et al. Genetic and environmental determinants of variation in the plasma lipidome of older Australian twins. eLife 2020;9:e58954 DOI: 10.7554/eLife.58954

7. Liu H, Wang W, Zhang C, Xu C, Duan H, Tian X and Zhang D (2018) Heritability and Genome-Wide Association Study of Plasma Cholesterol in Chinese Adult Twins. *Front. Endocrinol.* 9:677. doi: 10.3389/fendo.2018.00677

8. Zimoń, M., Huang, Y., Trasta, A. et al. Pairwise effects between lipid GWAS genes modulate lipid plasma levels and cellular uptake. Nat Commun 12, 6411 (2021). https://doi.org/10.1038/s41467-021-26761-3

9. Truong T-H, Do D-L, Kim N-T, Nguyen M-NT, Le T-T and Le H-A (2020) Genetics, Screening, and Treatment of Familial Hypercholesterolemia: Experience Gained From the Implementation of the Vietnam Familial Hypercholesterolemia Registry. *Front. Genet.* 11:914. doi: 10.3389/fgene.2020.00914

10. Vrablik M, Tichý L, Freiberger T, Blaha V, Satny M and Hubacek JA (2020) Genetics of Familial Hypercholesterolemia: New Insights. *Front. Genet.* 11:574474. doi: 10.3389/fgene.2020.574474
11. Guo Q, Feng X and Zhou Y (2020) *PCSK9* Variants in Familial Hypercholesterolemia: A Comprehensive Synopsis. *Front. Genet.* 11:1020. doi: 10.3389/fgene.2020.01020
12. Dušková L, Nohelová L, Loja T, Fialová J, Zapletalová P, Réblová K, Tichý L, Freiberger T and Fajkusová L (2020) Low Density Lipoprotein Receptor Variants in the Beta-Propeller Subdomain and Their Functional Impact. *Front. Genet.* 11:691. doi: 10.3389/fgene.2020.00691
13. Chasman DI, Posada D, Subrahmanyan L, Cook NR, Stanton, Jr VP, Ridker PM. Pharmacogenetic Study of Statin Therapy and Cholesterol Reduction. *JAMA.* 2004;291(23):2821–2827. doi:10.1001/jama.291.23.2821
14. Postmus, I., Trompet, S., Deshmukh, H. et al. Pharmacogenetic meta-analysis of genome-wide association studies of LDL cholesterol response to statins. Nat Commun 5, 5068 (2014). https://doi.org/10.1038/ncomms6068
15. Njelekela, M.A., Mpembeni, R., Muhihi, A. et al. Gender-related differences in the prevalence of cardiovascular disease risk factors and their correlates in urban Tanzania. BMC Cardiovasc Disord 9, 30 (2009). https://doi.org/10.1186/1471-2261-9-30
16. Giuseppina Russo, Basilio Pintaudi, Carlo Giorda, Giuseppe Lucisano, Antonio Nicolucci, Maria Rosaria Cristofaro, Concetta Suraci, Maria Franca Mulas, Angela Napoli, Maria Chiara Rossi, Valeria Manicardi, "Age- and Gender-Related Differences in LDL-Cholesterol Management in Outpatients with Type 2 Diabetes Mellitus", International Journal of Endocrinology, vol. 2015, Article ID 957105, 8 pages, 2015. https://doi.org/10.1155/2015/957105
17. Zhang, X., Ji, L., Ran, X. et al. Gender Disparities in Lipid Goal Attainment among Type 2 Diabetes Outpatients with Coronary Heart Disease: Results from the CCMR-3B Study. Sci Rep 7, 12648 (2017). https://doi.org/10.1038/s41598-017-13066-z
18. Wang, M., Liu, M., Li, F., Guo, C., Liu, Z., Pan, Y., Liu, Y., Liu, F., Cai, H., Wu, Y., He, Z., & Ke, Y. (2020). Gender heterogeneity in dyslipidemia prevalence, trends with age and associated factors in

middle age rural Chinese. *Lipids in health and disease, 19*(1), 135. https://doi.org/10.1186/s12944-020-01313-8

19. Tran, N., Blizzard, C. L., Luong, K. N., Truong, N., Tran, B. Q., Otahal, P., Nelson, M. R., Magnussen, C. G., Van Bui, T., Srikanth, V., Au, T. B., Ha, S. T., Phung, H. N., Tran, M. H., Callisaya, M., & Gall, S. (2021). Sex differences in total cholesterol of Vietnamese adults. *PloS one, 16*(8), e0256589. https://doi.org/10.1371/journal.pone.0256589
20. Karvinen S, Jergenson MJ, Hyvärinen M, Aukee P, Tammelin T, Sipilä S, Kovanen V, Kujala UM and Laakkonen EK (2019) Menopausal Status and Physical Activity Are Independently Associated With Cardiovascular Risk Factors of Healthy Middle-Aged Women: Cross-Sectional and Longitudinal Evidence. *Front. Endocrinol.* 10:589. doi: 10.3389/fendo.2019.00589
21. Saikia N, Moradhvaj, Bora JK (2016) Gender Difference in Health-Care Expenditure: Evidence from India Human Development Survey. PLoS ONE 11(7): e0158332. https://doi.org/10.1371/journal.pone.0158332
22. Spector ND, Overholser B. Examining Gender Disparity in Medicine and Setting a Course Forward. *JAMA Netw Open.* 2019;2(6):e196484. doi:10.1001/jamanetworkopen.2019.6484

CHAPTER 5

The Gut Microbiota And Cholesterol

Contents

5.1 BACKGROUND:

"We inherit every one of our genes, but we leave the womb without a single microbe. As we pass through our mother's birth canal, we begin to attract entire colonies of bacteria. By the time a child can crawl, he has been blanketed by an enormous, unseen cloud of microorganisms--a hundred trillion or more. They are bacteria, mostly, but also viruses and fungi (including a variety of yeasts), and they come at us from all directions: other people, food, furniture, clothing, cars, buildings, trees, pets, even the air we breathe. They congregate in our digestive systems and our mouths, fill the space between our teeth, cover our skin, and line our throats. We are inhabited by as many as ten thousand bacterial species; those cells outnumber those which we consider our own by ten to one, and weigh, all told, about three pounds--the same as our brain. Together, they are referred to as our microbiome--and they play such a crucial role in our lives that scientists like [Martin J.] Blaser have begun to reconsider what it means to be human."

— *Michael Specter*

The human gut hosts close to 10^{14} cells in the entire gut of a healthy individual. Co-evolution has facilitated colonization by these commensals to play a slew of roles in health (Li and Zhang, 2017; Shivsanjeevi Sripathi and Prerna Pandey, 2022).

The resident microbes are predominantly *Firmicutes, Bacteroidetes,* and *Actinobacteria* with the "microbiome" augmenting the host genome by a factor of 50 to 100. These added genes are vitally involved in controlling our metabolism entailing intricate symbiotic associations between the microbes and us. Many metabolites originating from the microbes are vitally involved in the maintenance of health (Kho and Lal, 2018).

Kho and Lal (2018) have presented a summary of gut microbe metabolites and their role that is depicted in Table 1 below: ***TOP***: Metabolites and functions (© 2018 Kho and Lal. *This is an open-access article distributed under the terms of the Creative Commons Attribution License (CC BY). The use, distribution or reproduction in other forums is permitted, provided the original author(s) and the copyright owner(s) are credited and that the original publication in this journal is cited, in accordance with accepted academic practice. No use, distribution or reproduction is permitted which does not comply with these terms*):

Metabolites	Functions	References
Short-chain fatty acids (SCFAs): E.g., Acetate, butyrate, propionate, hexanoate, valerate	• Regulate host metabolic pathways via G-protein-coupled receptor GPR41 or GPR43 -mediated signaling: • energy homeostasis; synthesis of glucagon-like peptide 1 (GLP-1); increase leptin production. • Improve glucose tolerance and insulin sensitivity. • Potent histone deacetylase (HDAC) inhibitor - regulation of intestinal cell proliferation. • Intestinal gluconeogenesis, lipogenesis, suppression of fasting-induced adipose factor Fiaf (lipoprotein lipase inhibitor) in intestinal epithelium. • Immunomodulatory effect, activate dendritic cells, gut immunity.	Blottière et al., 2003; Backhed et al., 2004; Xiong et al., 2004; Samuel et al., 2008; Tolhurst et al., 2012; De Vadder et al., 2014
Indole derivatives: E.g., Indole, indoxyl sulfate, indole-3-propionic acid (IPA)	• IPA as powerful antioxidant, inhibitor of amyloid-beta fibril formation, and exhibits neuroprotective and cytoprotective effects against a variety of oxidotoxins. • IPA regulates intestinal barrier function via the xenobiotic sensor, pregnane X receptor (PXR), in which it reduces intestinal inflammation (downregulates enterocyte pro-inflammatory cytokines TNF-α), and regulate intestinal permeability and mucosal integrity (upregulates junctional protein-coding mRNAs). • Indoxyl sulfate as uremic toxin that accumulates in the blood of patients with impaired excretion system.	Bendheim et al., 2002; Deguchi et al., 2002; Wikoff et al., 2009; Venkatesh et al., 2014
Bile acid metabolites: E.g., Deoxycholic acid (DCA), lithocholic acid (LCA)	• Activate host nuclear receptors and cell signaling pathways: regulation of bile acid, cholesterol, glucose, lipid, and energy metabolism. • Exhibit antimicrobial effects.	Inagaki et al., 2006; Hylemon et al., 2009
Choline metabolites: E.g., Choline, trimethylamine N-oxide (TMAO) and betaine	• Modulate lipid metabolism and glucose homeostasis. • Contribute to non-alcoholic fatty liver disease and cardiovascular disease.	Dumas et al., 2006; Wang Z. et al., 2011
Phenolic derivatives: E.g., 4-OH phenylacetic acid, equol, urolithins, enterolactone, enterodiol, 8-prenylnaringenin, 2-(3,4-dihydroxyphenyl)acetic acid, 3-(4-hydroxyphenyl)propionic acid, 5-(3,4-dihydroxyphenyl)valeric acid	• Antimicrobial effects: repress pathogenic microbes, influence gut microbiota composition, maintenance of intestinal health. • Protective effect against oxidative stress. • Estrogen-modulating effect. • Platelet aggregation inhibition effect. • Urolithin exhibits anti-inflammatory and cancer chemopreventive effects.	Larrosa et al., 2006, 2010; Lee H.C. et al., 2006; Selma et al., 2009
Vitamins: E.g., Thiamine (B1), riboflavin (B2), niacin (B3), pyridoxine (B6), pantothenic acid (B5), biotin (B7), folate (B11–B9), cobalamin (B12), and menaquinone (K2)	• Energy production, red blood cell formation, as enzymatic cofactor for diverse biochemical reactions. • DNA replication, repair and methylation, regulating cell proliferation.	LeBlanc et al., 2011; Nicholson et al., 2012; Forster et al., 2017; Lerner et al., 2017b
Polyamines: E.g., putrescine, spermidine, and spermine	• Production of nucleotides, vitamins and amino acids. • Enhance immune functioning. • Sustain high proliferation rate of Intestinal epithelial cells. • Dysregulated polyamine metabolism possibly enhances cancer development. • Enhance intestinal barrier integrity and function via stimulating synthesis of intercellular junction proteins • [occludin, zonula occludens-1 (ZO-1), E-cadherin]. • Enhance maturation of intestinal and systemic adaptive immune system. • Spermine inhibits pro-inflammatory M1 macrophage activation.	Guo et al., 2003, 2005; Perez-Cano et al., 2010; Johnson et al., 2015; Rooks and Garrett, 2016

BOTTOM: Dysbiosis in gut bacteria and few associated maladies (© 2018 Kho and Lal. *This is an open-access article distributed under the terms of the Creative Commons Attribution License (CC BY). The use, distribution or reproduction in other forums is permitted, provided the original author(s) and the copyright owner(s) are credited and that the original publication in this journal is cited, in accordance with accepted academic practice. No* *use, distribution or reproduction is permitted which does not comply with these terms*):

Disease categories	Specific diseases	Associated dysbiotic features	Reference
Immune-mediated/autoimmune diseases	inflammatory bowel disease (IBD)	• Increase in virulent gut microbes (*Enterobacteriaceae* species, *Bacteroides fragilis*) and mucolytic *Ruminococcus* sp. • Decrease in butyrate-producing *Firmicutes* (such as *Faecalibacterium prausnitzii*, *Roseburia hominis*)	Sokol et al., 2008; Png et al., 2010; Willing et al., 2010; Machiels et al., 2014
	Irritable bowel syndrome (IBS)	• Increase in *Escherichia coli* • Decrease in *Clostridium leptum* group of bacteria and *Bifidobacterium*. • Decrease in bile acid biotransformation	Duboc et al., 2012
	Celiac disease	• Increase in *Bacteroides–Prevotella* group • Decrease in *Bifidobacterium* • Varying observation (decrease or no change) in *Clostridium histolyticum*, *C. lituseburense*, and *Faecalibacterium prausnitzii* • Alteration in SCFAs composition, but overall increase in total SCFA	Tjellström et al., 2005; Nadal et al., 2007; De Palma et al., 2010
	Systemic lupus erythematosus (SLE)	• Increase in *Blautia* sp. and Gram-negative bacteria such as *Proteobacteria*. • Decrease in gut microbiota diversity, *Odoribacter* sp., *Alistipes* sp. • Increase in serum endotoxin	Shi et al., 2014; Luo et al., 2018
	Type-1 diabetes	• Increase in *Bacteroidetes* • Decrease in *Actinobacteria*, *Firmicutes*, and *Firmicutes/Bacteroidetes* ratio	Murri et al., 2013
	Rheumatoid arthritis (RA)	• Increase in *Prevotella copri* and decrease in *Bacteroides* sp. In new-onset RA • Increase in microbiota diversity of *Lactobacillus* genus in early RA	Liu et al., 2013; Schar et al., 2013
	Atopic disease (E.g., childhood allergic asthma)	• Increase in fecal burden of *Clostridium difficile*, and *C. difficile/Bifidobacteria* ratio	Kalliomäki et al., 2001
Metabolic disorders/cardiovascular disorders	Obesity	• Increase in *Firmicutes, Actinobacter* • Varying observation (decrease, no change, increase) in *Bacteroidetes* • Increase in glycoside hydrolase and SCFAs (butyrate and acetate)	Turnbaugh et al., 2006, 2009a; Koliada et al., 2017
	Type-2 diabetes	• Increase in *Lactobacillus* • Decrease in *Clostridium coccoides*, *Atopobium* cluster, and *Prevotella* • Decrease in butyrate biosynthesis	Qin et al., 2012; Sato et al., 2014
	Hypertension	• Increase in the Firmicutes/Bacteroidetes ratio, lactate-producer • Decrease in microbiota diversity, acetate- and butyrate-producers	Yang et al., 2015
	Atherosclerosis	• Increase in metabolites TMAO, endotoxin level (risk factor for early atherosclerosis)	Wiedermann et al., 1999; Koeth et al., 2013
Cancer	Colorectal cancer (CRC)	• Increase in enterotoxigenic *Bacteroides fragilis*, and pathobionts *Fusobacterium* and *Campylobacter* sp. • Decrease in butyrate-producer (*Faecalibacterium* and *Roseburia*)	Wang et al., 2012; Wu et al., 2013
Neurop sy chi atri c	Autism spectrum disorder (ASD)	• Increase in *Clostridium* sp., *Bacteroidetes*, *Lactobacillus*, *Desulfovibrio* • Decrease in *Bifidobacteria*	Song et al., 2004; Adams et al., 2011
	Alzheimer's disease	• Possible connection between gut microbiota-synthesized amyloids, LPS, γ-aminobutyric acid (GAB A – major inhibitory neurotransmitter), and the increased permeability of gut barrier and blood brain barrier with age	Pistollato et al., 2016
	Depression	• Increase in genus *Eggerthella*, *Holdemania*, *Gelria*, *Turicibacter*, *Paraprevotella*, *Anaerofilum* • Decrease in gut microbiota diversity, *Prevotella* and *Dialister*	Kelly et al., 2016
	Parkinson's Disease	• Increase in anti-inflammatory butyrate-producers from genus *Blautia*, *Coprococcus*, and *Roseburia* in patient fecal sample, pro-inflammatory *Proteobacteria* in patient mucosa • Increased gene expression in LPS biosynthesis and microbial type III secretion system	Keshavarzian et al., 2015
Infectious disease	*Clostridium difficile* infection (CDI)	• Increase in *Clostridium difficile*; • Decrease in gut microbiota diversity and secondary bile acids-producing *Clostridium scindens*	Theriot et al., 2014
Uremic disease	Chronic kidney disease	• Increase in *Firmicutes*, *Proteobacteria*, and *Actinobacteria* • Decrease in *Lactobacilli*	Vaziri et al., 2013

Wong and team proposed a model where the lipidome is influenced by genetics and environmental influences as depicted below in figure 1 (Matthew WK Wong, Anbupalam Thalamuthu, Nady Braidy et al, 2020 © 2020, Wong et al. *This article is distributed under the terms of the Creative Commons Attribution License, which permits unrestricted use and redistribution provided that the original author and source are credited*):

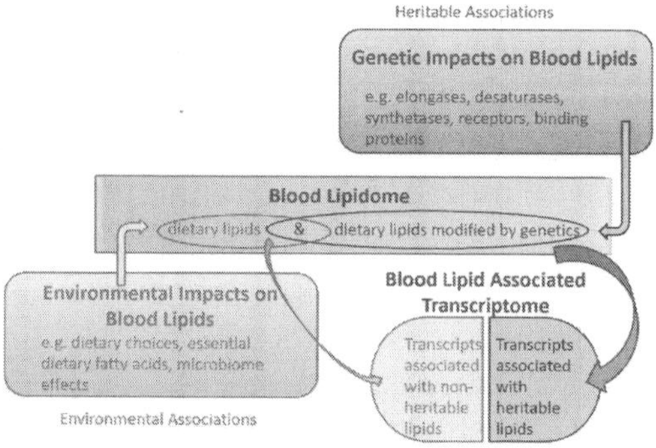

5.2 FEW RESEARCH STUDIES

In order to probe the impact of cholesterol on gut microbes, Bo, Shao, Wu, Niu, Zhao and Gao (2017) reported high-throughput sequencing to scrutinize the microbiome of wild-type Sprague Dawley (SD) rats and ApoE $^{-/-}$ (Apolipoprotein E or ApoE that is vitally involved in lipoprotein biology and cholesterol metabolism) SD rats administered normal chow and high-cholesterol diets. The categorization of ApoE$^{-/-}$ and wild-type (WT) animals was normal chow (NC) and high-cholesterol (HC) diets followed by analyses of cholesterol (TC), triglycerides (TG), LDL, and HDL levels in the serum. Figure 2 below illustrates (a) cholesterol profile and (b) body weights of the animals as reported by the researchers (Bo, Shao, Wu, Niu, Zhao and Gao, 2017: © 2017 The Authors. Microbiology Open published by John Wiley & Sons Ltd. *This is an open access article under the terms of the Creative Commons Attribution License, which permits use, distribution and reproduction in any medium, provided the original work is properly cited*):

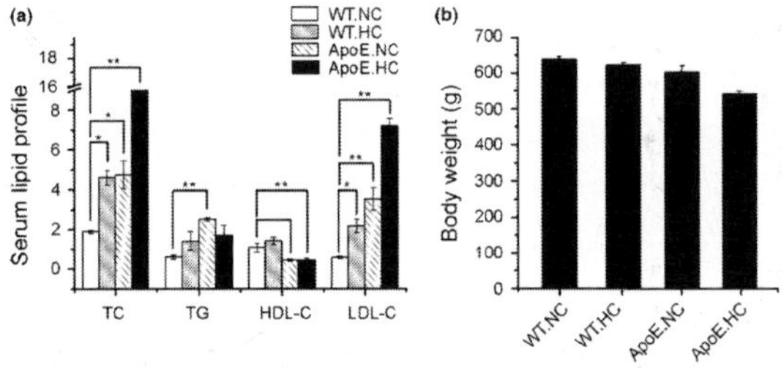

Analyses of the Shannon index revealed that the diversity of the gut microbiota was diminished by the high-cholesterol diet as opposed to the WT. The HC group also demonstrated a lowered Firmicutes to Bacteroidetes (F/B) ratio indicative of altered microbiota due to the cholesterol spike. The HC group also demonstrated increased *Prevotella* (seen in obesity), *Euryarchaeota* (methanogens implicated to generate toxic intermediates) and *Spirochaetes* (known pathogens). Notably, the ApoE groups documented a conspicuous lowering of the microbial alterations from the dietary changes that is indicative of the plausible role of genetics to influence dietary impacts.

Figure 3 below depicts the findings of the researchers: ***TOP***: The bacterial phyla and their relative abundance (Bo, Shao, Wu, Niu, Zhao and Gao, 2017: © 2017 The Authors. Microbiology Open published by John Wiley & Sons Ltd. *This is an open access article under the terms of the Creative Commons Attribution License, which permits use, distribution and reproduction in any medium, provided the original work is properly cited*):

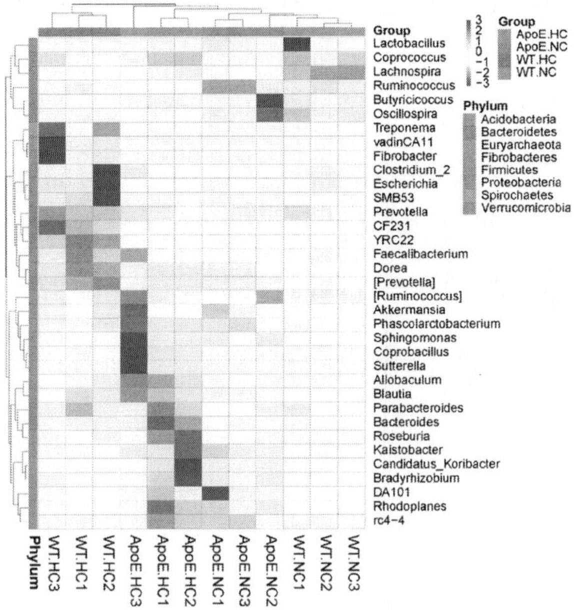

BOTTOM: Relative abundance of specific organisms: (a) Firmicutes and Bacteroidetes; (b) F/B ratio (c) Euryarchaeota, Spirochaetes, Fibrobacteres (d) *Prevotella, Oscillospira, Ruminococcus, Bacteriodes* (e) other species (f) ApoE group and other species: $*p < .05$; $\#p < .05$; $\Delta p < .05$ (Bo, Shao, Wu, Niu, Zhao and Gao, 2017: © 2017 The Authors. Microbiology Open published by John Wiley & Sons Ltd. *This is an open access article under the terms of the Creative Commons Attribution License, which permits use, distribution and reproduction in any medium, provided the original work is properly cited*):

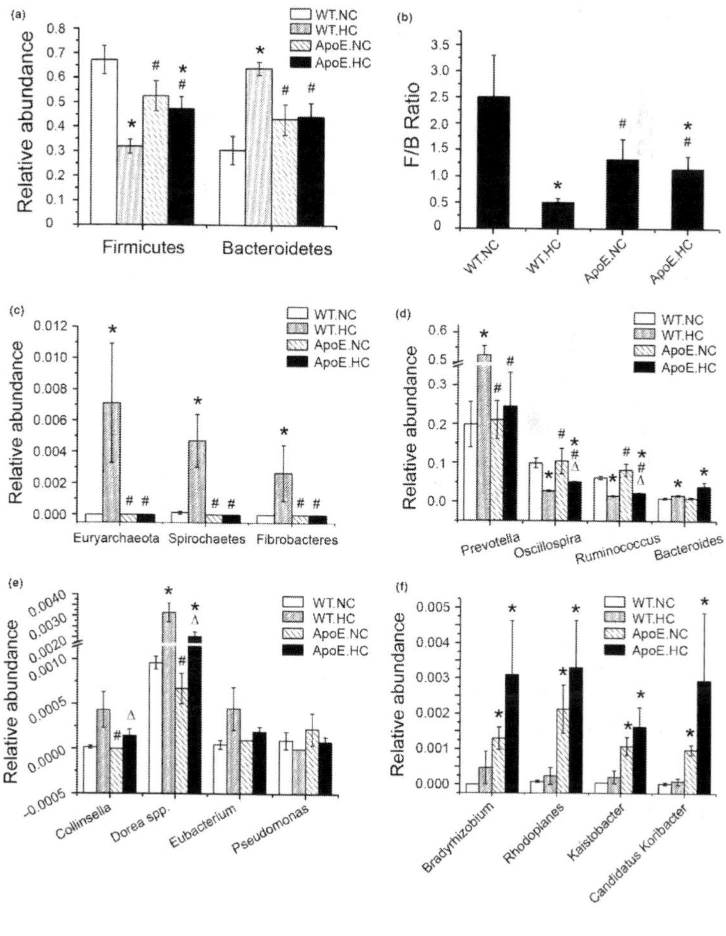

The impact of gut microbiota on cholesterol homeostasis was scrutinized by Le Roy, Lécuyer, Chassaing et al (2019) by employing vancomycin, ampicillin, neomycin, and metronidazole in hypercholesterolemic *Apoe*$^{-/-}$ mice to diminish their intestinal microbiota. These microbiota-depleted (AB-Mdpl) mice documented a spike in plasma total cholesterol by 55% as opposed to conventionally raised (Conv-R) mice. Analyses of chylomicrons, very-low-density lipoproteins (VLDL), low-density lipoproteins (LDL), and high-density lipoproteins (HDL) unveiled similar HDL profiles in AB-Mdpl and Conv-R animals while the VLDL and LDL documented an increase in the former group. A gavage of [^{3}H]-cholesterol in olive oil revealed that the AB-Mdpl mice documented a rise in plasma cholesterol in the blood by 25% vs. the Conv-R ones. Jejunum gene analyses revealed alterations in genes functioning in *Npc1l1* (cholesterol absorption) and *Abcg8* (cholesterol excretion).

Figure 4 below depicts the findings of altered cholesterol profiles due to lowered intestinal microbiota as reported by the researchers (Le Roy, Lécuyer, Chassaing et al, 2019: Open Access: *This article is distributed under the terms of the Creative Commons Attribution 4.0 International License (http://creativecommons.org/licenses/by/4.0/), which permits unrestricted use, distribution, and reproduction in any medium, provided you give appropriate credit to the original author(s) and the source, provide a link to the Creative Commons license, and indicate if changes were made. The Creative Commons Public Domain Dedication waiver (http://creativecommons.org/publicdomain/zero/1.0/) applies to the data made available in this article, unless otherwise stated*):

a The design **b** Plasma levels of cholesterol, phospholipids and triglycerides in conventionally raised (Conv-R) vs. microbiota-depleted mice (AB-Mdpl). **c** Fast protein liquid chromatography for cholesterol distribution in VLDL, LDL and HDL lipoproteins **d** [^{3}H]-cholesterol and plasma radioactivity **e** Relative expression of cholesterol absorption genes in the jejunum. **f** Relative expression of lipoprotein genes in the jejunum:

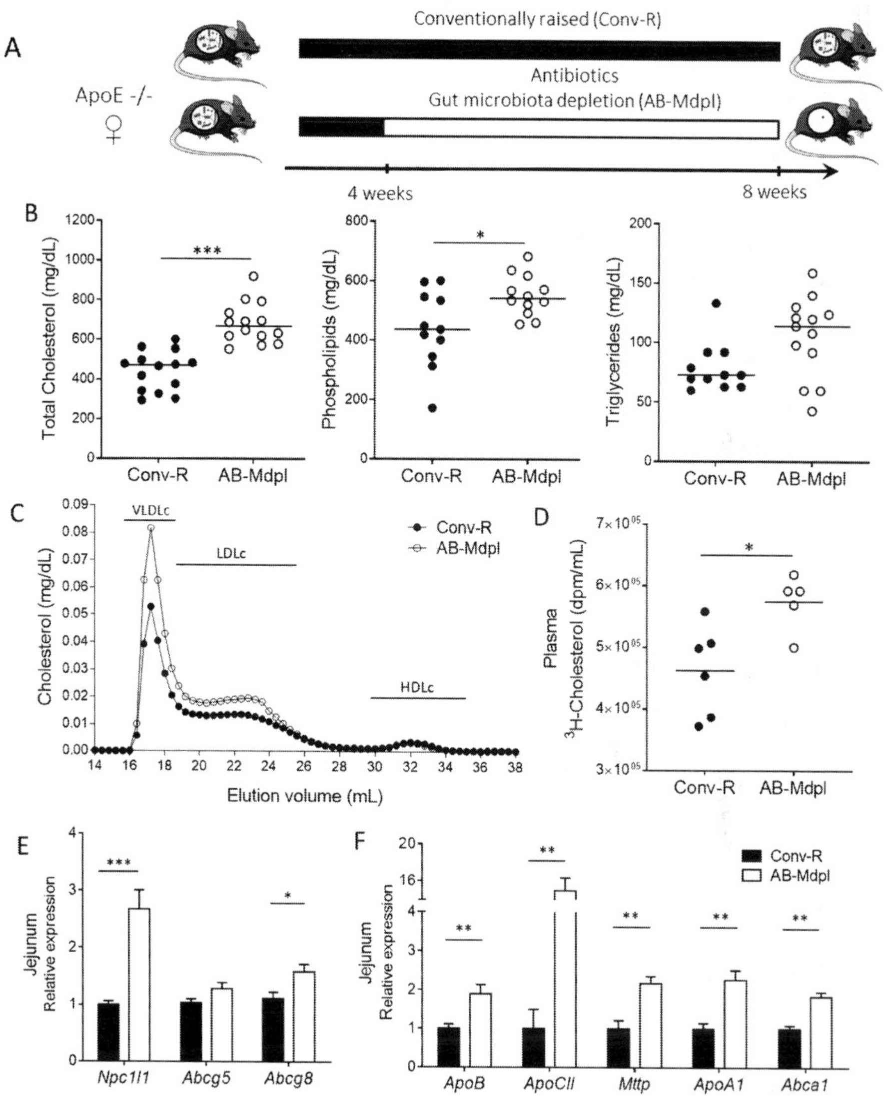
A
Conventionally raised (Conv-R)
ApoE -/-
♀
Antibiotics
Gut microbiota depletion (AB-Mdpl)
4 weeks
8 weeks
B
Total Cholesterol (mg/dL)
***
Conv-R
AB-Mdpl
Phospholipids (mg/dL)
*
Triglycerides (mg/dL)
C
VLDLc
LDLc
HDLc
Cholesterol (mg/dL)
Elution volume (mL)
Conv-R
AB-Mdpl
D
Plasma
^{3}H-Cholesterol (dpm/mL)
E
Jejunum
Relative expression
Npc1l1
Abcg5
Abcg8
F
ApoB
ApoCII
Mttp
ApoA1
Abca1

The microbiota-depleted mice documented increased LDL receptor (LDLr) mRNA expression (functioning in liver uptake of LDL) while no change was documented for scavenger receptor type B1 (SR-B1) (involved in HDL uptake). Analyses in $LDLr^{-/-}$ mice revealed a 91% rise in plasma cholesterol in LDLr-deficient mice as opposed to a rise of 50% in Apoe-deficient mice suggestive of the partial counteraction of LDLr-mediated cholesterol uptake by the liver of the rise in plasma cholesterol due to the gut microbiota depletion. AB-Mdpl mice also documented an elevation of 40% in the bile flow with increased biliary cholesterol secretion in the intestinal lumen. Further, ^{14}Ccholesterol-based studies revealed a lowering of cholesterol by a factor of three and a rise in bile acids by 50% in AB-Mdpl mice feces indicative of sterol accumulation due to gut microbiota depletion. These animals also demonstrated a lowering of *Fgf-15* by 75%- this becomes interesting as FXR from gut microbiota controls the expression of the rate-limiting enzyme in the bile acid synthesis pathway (Cyp7a1) via fibroblast growth factor 15 (Fgf-15).

The alterations in the cholesterol pathway gene profiles due to gut microbiota depletion in the mice models are depicted in figure 5 below: ***TOP***: **a** [^{3}H]-cholesterol gavage and radioactivity **b** Relative expression of cholesterol transporters in the liver **c** Rise in cholesterol in microbiota-depleted mice vs. Apo e (○) and LDLr (□)$^{-/-}$ mice. **d** Cholesterol synthesis gene expression in the liver **e** GC-MS off cholesterol and lathosterol in the liver (Le Roy, Lécuyer, Chassaing et al, 2019: Open Access: *This article is distributed under the terms of the Creative Commons Attribution 4.0 International License (http://creativecommons.org/licenses/by/4.0/), which permits unrestricted use, distribution, and reproduction in any medium, provided you give appropriate credit to the original author(s) and the source, provide a link to the Creative Commons license, and indicate if changes were made. The Creative Commons Public Domain Dedication waiver (http://creativecommons.org/publicdomain/zero/1.0/) applies to the data made available in this article, unless otherwise stated*):

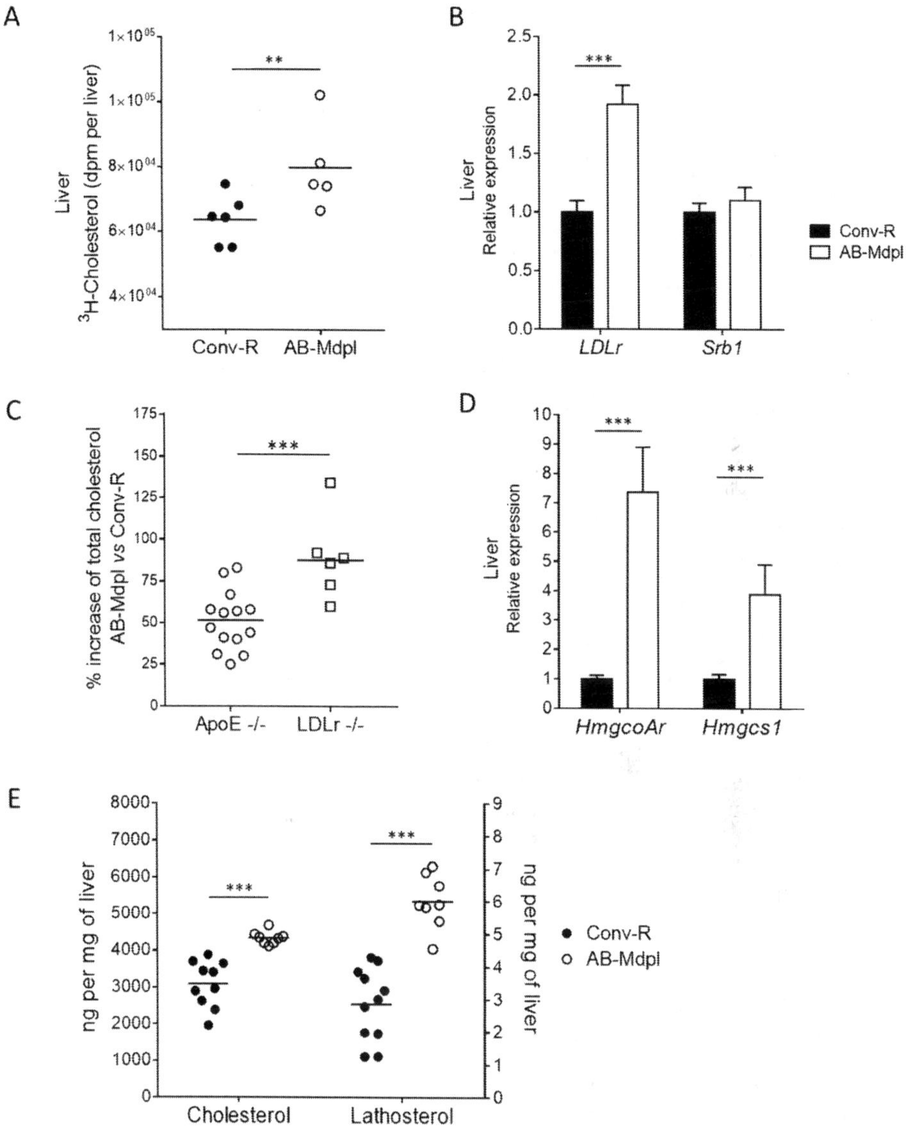

BOTTOM: **a** Bile volume analyses **b** Cholesterol secretion in the bile during gallbladder cannulation (1h) **c** Bile acid biosynthesis, cholesterol and bile acid gene profiles **d** ^{14}C bile acids (water-soluble fraction) and ^{14}C cholesterol (cyclohexane soluble fraction) excretion following ^{14}C cholesterol gavage **e** Excretion of ^{14}C bile acids **f** *fgf15* expression in the distal ileum. **g** [^{3}H]-taurocholic acid gavage **h** Bile acid transporter expression in the distal ileum (Le Roy, Lécuyer, Chassaing et al, 2019: Open Access: *This article is distributed under the terms of the Creative Commons*

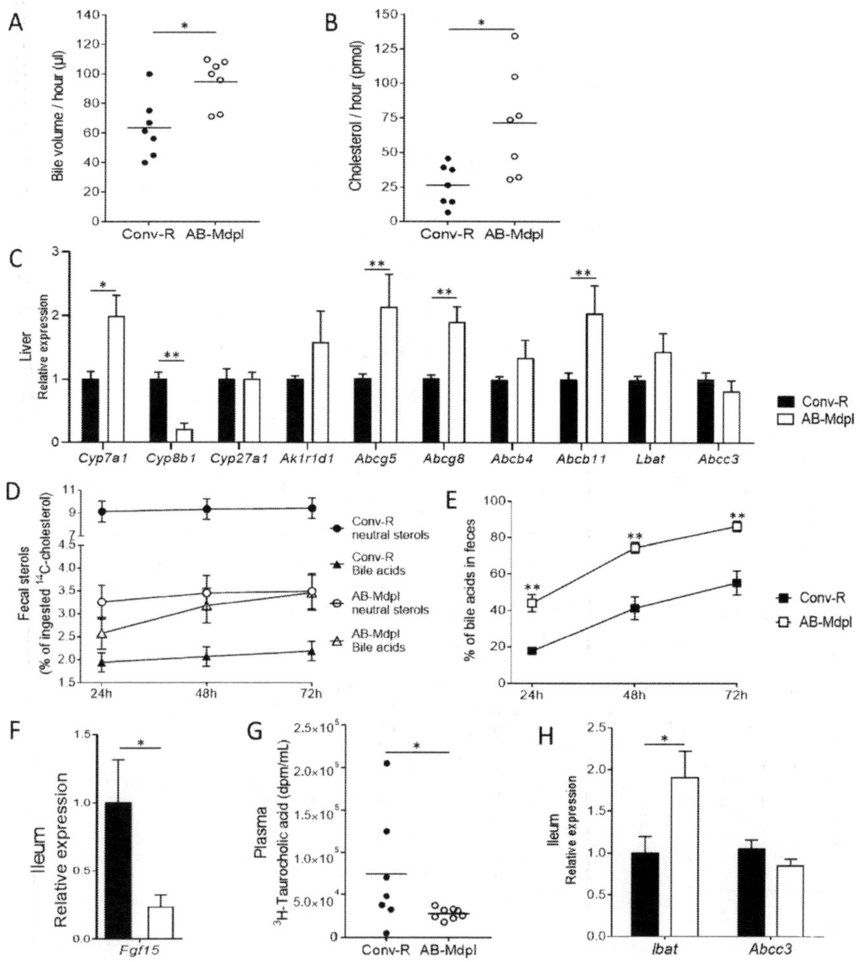
A
Bile volume / hour (µl)
Conv-R AB-Mdpl
B
Cholesterol / hour (pmol)
Conv-R AB-Mdpl
C
Liver
Relative expression
Cyp7a1 Cyp8b1 Cyp27a1 Ak1r1d1 Abcg5 Abcg8 Abcb4 Abcb11 Lbat Abcc3
Conv-R
AB-Mdpl
D
Fecal sterols
(% of ingested ¹⁴C-cholesterol)
Conv-R neutral sterols
Conv-R Bile acids
AB-Mdpl neutral sterols
AB-Mdpl Bile acids
24h 48h 72h
E
% of bile acids in feces
Conv-R
AB-Mdpl
24h 48h 72h
F
Ileum
Relative expression
Fgf15
G
Plasma
³H-Taurocholic acid (dpm/mL)
Conv-R AB-Mdpl
H
Ileum
Relative expression
Ibat
Abcc3

Transplanting the microbiota from donors with normal blood cholesterol levels (NorChol) and donors with moderately elevated total cholesterol levels (HiChol) in microbiota-depleted 7-week-old female *Apoe*$^{-/-}$ mice revealed an evident increase of plasma total cholesterol levels of HiChol recipients vs. the NorChol recipient mice documenting the transmissibility of plasma cholesterol levels via the gut microbiota. The high cholesterol groups also demonstrated boosted *Npc1l1*, *ApoB*, *ApoCII*, and *Mtpp* expression cholesterol absorption and lipoprotein secretion), lowered *HmgcoAr* and *Hmgcs1* (cholesterol synthesis) and also reduced LDL receptor (reduced uptake) vs. mice recipients with regular cholesterol level donor microbiota.

Subsequent 16S rRNA sequencing unveiled that the HiChol recipients documented an abundance of ten ASVs or amplicon sequence variants that documented correlations with the plasma cholesterol levels paving the way for further elucidation of cholesterol absorption and microbiota.

Figure 6 below illustrates the findings of the researchers: ***TOP***: intestinal microbiota **a** Interclass principal component analysis **b** Cladogram where NorChol recipients and HiChol recipients are highlighted in cyan and red, respectively for altered abundances **c** Statistical abundance of ASVs **d** ASV-level microbial populations and cholesterol metabolism-associated parameters correlation with large circles for strong correlations are indicated by large circles and small circles for weaker correlations; dark blue-strong positive correlation; dark red-strong negative correlation (Le Roy, Lécuyer, Chassaing et al, 2019: Open Access: *This article is distributed under the terms of the Creative Commons Attribution 4.0 International License (http://creativecommons.org/licenses/by/4.0/), which permits unrestricted use, distribution, and reproduction in any medium, provided you give appropriate credit to the original author(s) and the source, provide a link to the Creative Commons license, and indicate if changes were made. The Creative Commons Public Domain Dedication waiver (http://creativecommons.org/publicdomain/zero/1.0/) applies to the data made available in this article, unless otherwise stated*):

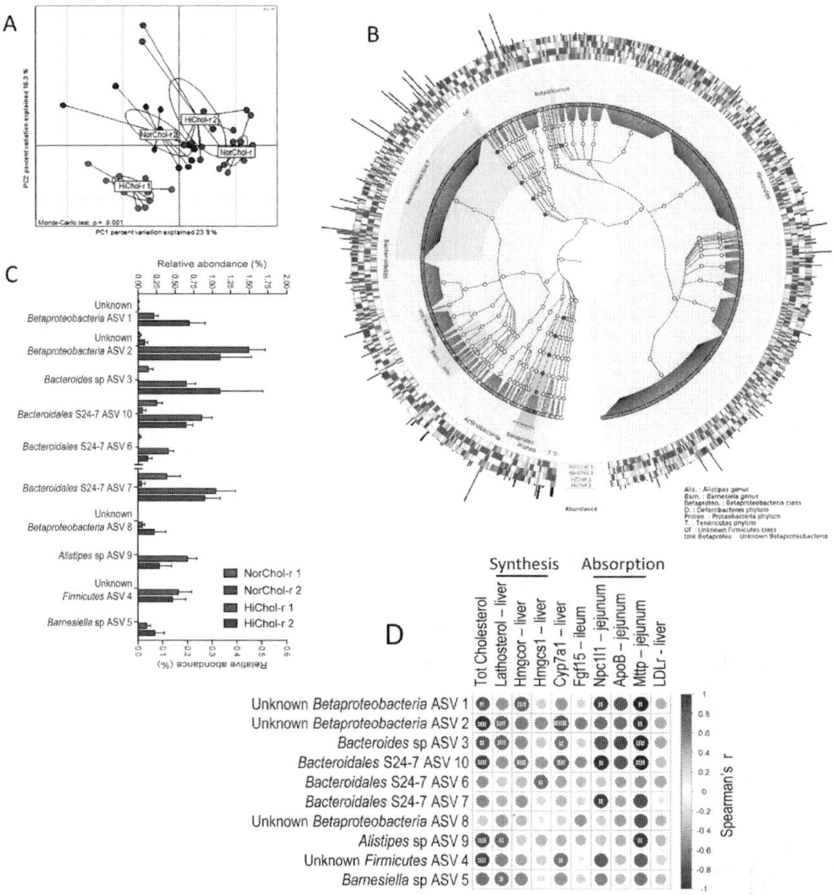

BOTTOM: A model of how hepatic cholesterol and bile acid synthesis, hepatic cholesterol uptake and intestinal cholesterol absorption are elevated in microbiota-depleted mice due to lowered *fgf15* expression in the distal ileum (Le Roy, Lécuyer, Chassaing et al, 2019: Open Access: *This article is distributed under the terms of the Creative Commons Attribution 4.0 International License (http://creativecommons.org/licenses/by/4.0/), which permits unrestricted use, distribution, and reproduction in any medium, provided you give appropriate credit to the original author(s) and the source, provide a link to the Creative Commons license, and indicate if changes were made. The Creative Commons Public Domain Dedication waiver (http://creativecommons.org/publicdomain/zero/1.0/) applies to the data made available in this article, unless otherwise stated*):

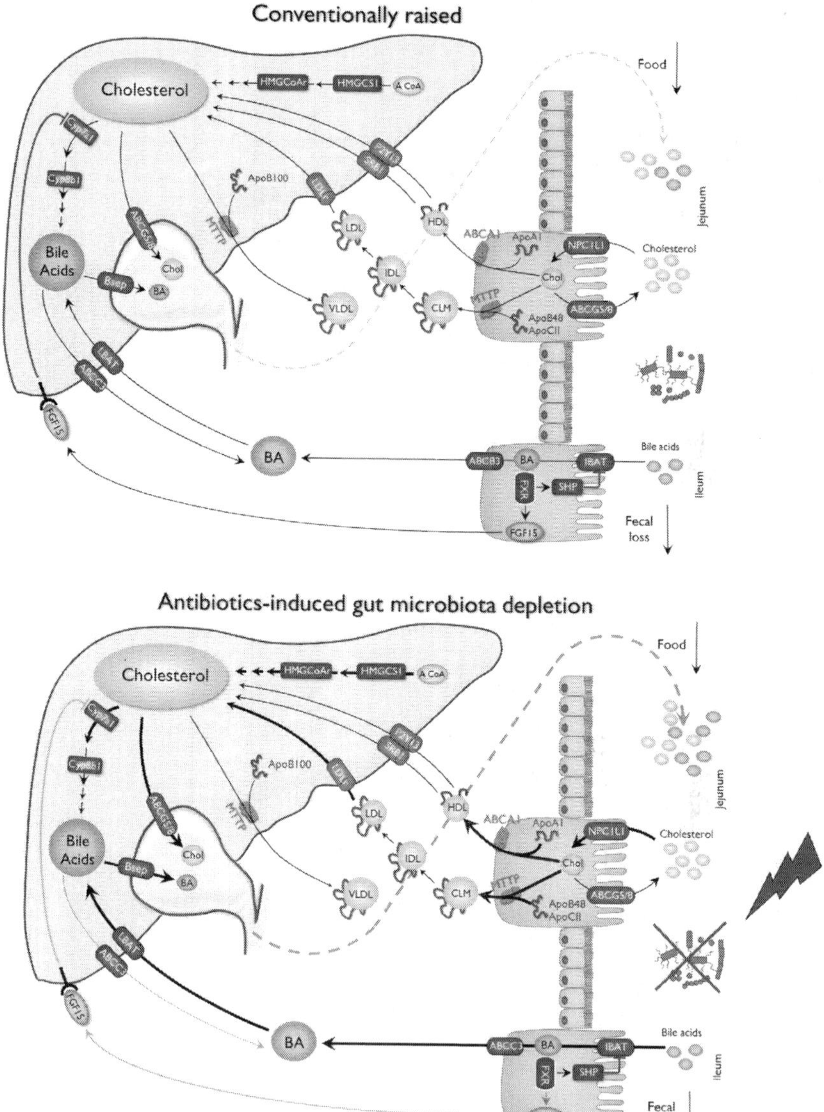

Other preclinical studies probing the link between gut microbiota and cholesterol homeostasis have been summarized by Villette, KC, Beliard, Salas Tapia, Rainteau, Guerin and Lesnik, *Frontiers In Pharmacology* in table 2 below (© 2020 Villette, KC, Beliard, Salas Tapia, Rainteau, Guerin and Lesnik. *This is an open-access article distributed under the terms of the Creative Commons Attribution License (CC BY). The use, distribution or reproduction in other forums is permitted, provided the original author(s) and the copyright owner(s) are credited and that the original publication*

in this journal is cited, in accordance with accepted academic practice. No use, distribution or reproduction is permitted which does not comply with these terms):

Model	Diet	Phenotype in germ-free	References
Axenic Normolipidemic Rat	CD	Chol ↗	(Danielsson and Gustafsson, 1959)
	CD with 0.5% cholesterol	Chol →	(Danielsson and Gustafsson, 1959)
Axenic Normolipidemic Mice	HFD 0.03% cholesterol	Chol ↘	(Rabot et al., 2010)
	CD*	Chol ↘	(Velagapudi et al., 2010)
	CD	Chol →	(Sayin et al., 2013)
	WD with 0.2% cholesterol	Chol ↘	(Zhong et al., 2015)
	CD source of fat: lard	Chol ↗	(Caesar et al., 2016)
	CD source of fat: fish oil	Chol →	(Caesar et al., 2016)
	CD	Chol →	(Mistry et al., 2017)
Normolipidemic Mice (axenization by a mixture of antibiotics)	CD*	Chol ↘	(Joyce et al., 2014)
	CD	Chol →	(Out et al., 2015)
	CD	Chol →	(Zarrinpar et al., 2018)
	CD	Chol →	Personal observations
Axenic Dyslipidemic Mice	$Apoe^{-/-}$ 0.15% cholesterol diet	Chol →	Wright et al., JEM 2000
	$Apoe^{-/-}$ CD	Chol ↗	(Stepankova et al., 2010)
	$Apoe^{-/-}$ 2% cholesterol	Chol ↗	(Stepankova et al., 2010)
	$Apoe^{-/-}$ CD	Chol ↗	(Kasahara et al., 2017)
	$Apoe^{-/-}$ CD	Chol ↗	(Lindskog Jonsson et al., 2018)
	$Apoe^{-/-}$ WD	Chol →	(Lindskog Jonsson et al., 2018)
	$Ldlr^{-/-}$ CD	Chol ↗	(Kiouptsi et al., 2019)
	$Ldlr^{-/-}$ 0.2% cholesterol diet	Chol →	(Kiouptsi et al., 2019)
	$Apoe^{-/-}$ CD	Chol ↗	(Chen et al., 2016)
Dyslipidemic Mice (axenization by a mixture of antibiotics)	$Apoe^{-/-}$ 0.15% cholesterol diet	Chol ↗	(Chen et al., 2016)
	$Apoe^{-/-}$ CD	Chol ↗	(Le Roy et al., 2019)
	$Ldlr^{-/-}$ CD	Chol ↗	(Le Roy et al., 2019)
Dyslipidemic Mice FMT	$Apoe^{-/-}$ CD	Cholesterol levels transmitted**	(Le Roy et al., 2019)

*All studies involved mice of C57Bl6 genetic background, except for two studies that used Swiss Webster mice (indicated by *). CD, chow diet; HFD, high-fat diet; WD, western diet; FMT, fecal material transfer.*

***Phenotype in recipient mice.*

↗Increased; ↘decreased; →no changes.

5.3 GLIMPSES OF RESEARCH ON HUMAN GUT MICROBIOTA AND CHOLESTEROL METABOLISM:

Researchers de Moraes and team (2017) probed the metagenomics of fecal microbiota (16SrRNA sequencing, V4 region) in 268 participants from the ADVENTO—Analysis of Diet and Lifestyle for Cardiovascular Prevention in Seventh-Day Adventists study. This was followed by the categorization of the participants into *Bacteroides*, *Prevotella*, and *Ruminococcaceae* enterotypes. Those assigned the *Prevotella* enterotype documented more strict vegetarians and decreased LDL-C levels as opposed to the *Bacteroides* enterotype. The LDL-C levels were also reduced in strict vegetarians and lacto-ovo-vegetarians vs. omnivores within the *Prevotella* enterotype. Further, HDL-c was more in the strict vegetarians of the *Ruminococcaceae* enterotype. Looking at specific genera or species, a worse profile emerged for cardiometabolic risk factors for "*Streptococcus* (r systolic BP = 0.44, $p = 0.004$, r diastolic BP = 0.51, $p < 0.001$, r insulin = 0.33, $p = 0.04$, and r LDL-c = 0.40, $p = 0.009$) and *Desulfovibrio* (r BMI = 0.42, $p = 0.006$, r diastolic BP = 0.37, $p = 0.02$, r insulin = 0.32, $p = 0.04$, and r LDL-c = 0.40, $p = 0.01$)" for the *Prevotella* enterotype. In the *Ruminococcaceae* enterotype, a better cardiovascular profile emerged for "*Bifidobacterium* (r total cholesterol = −0.21, $p = 0.01$) and *Haemophilus* (r diastolic BP = −0.23, $p < 0.01$)" while aberrant lipid profiles were documented for *Blautia* "(HDL-c levels lowered: $r = -0.20$, $p = 0.02$) and *Roseburia* (r LDL-c = 0.24, $p < 0.01$ and r HDL-c = −0.28, $p < 0.001$)" (de Moraes et al, 2017).

Figure 7 below illustrates correlation heatmaps of gut microbiota genera/species and cardiovascular risk factors inclusive of cholesterol as documented by the researchers (de Moraes, Fernandes, da Silva, Almeida-Pititto, Gomes, Pereira and Ferreira, 2017:):

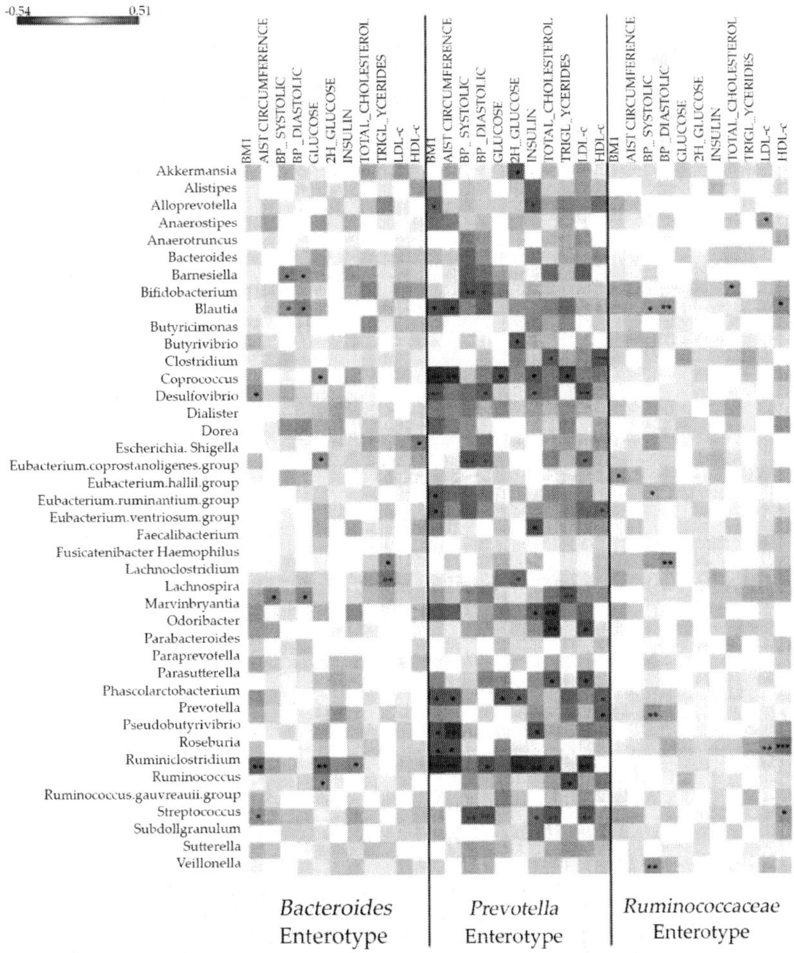

Zwartjes, Gerdes and Nieuwdorp (2021) reviewed alterations in gut microbiota in dyslipidemia as summarized by them in table 3 below (Zwartjes, Gerdes and Nieuwdorp, 2021 © 2021 by the authors. Licensee MDPI, Basel, Switzerland. *This article is an open access article distributed under the terms and conditions of the Creative Commons Attribution (CC BY) license: https://creativecommons.org/licenses/by/4.0/*):

Reference	Study Description	Subjects	Methods	Main Findings
Fu et al., 2015 [70]	Large cohort study	893 participants of the LifeLines-DEEP cohort	Lipid panels of total cholesterol, HDL-C, TGs, LDL-C. Sample genotyping. Fecal sample gut microbiota profiling with 16s rRNA gene sequencing. Operational taxonomic unit (OTU) picking and OTU richness. Microbial Shannon diversity index calculation. Cross validation analysis.	In total, 34 bacterial taxa correlated with plasma lipid levels and BMI. Specifically, family Clostridiaceae/ Lachnospiracease was correlated with LDL. Coprococcus (Firmicutes) and Collinsella had strong correlation with TGs. Cross validation analysis revealed that GM explained ≤25.9% of HDL-C variance and that microbiota explained 6% of the variance in TGs.
Rebolledo et al., 2017 [71]	Case control study	30 hypercholesterolemic subjects, 27 controls with normal cholesterol levels	Anthropometric data. Serum fasting glucose and lipid profile. Stool sample analysis of GM by gel electrophoresis with denaturing gradient (DGGE) technique and nonmetric multidimensional scaling (NMDS). Shannon–Weaver, Simpson, and Richness microbial community diversity indices.	DGGE banding profiles differed between case and control, confirmed by 2 separate groups forming on NMDS scaling analysis. Significant ($p < 0.05$) decrease of all three bacterial DNA indices of microbial diversity.
Vojinovic et al., 2019 [72]	Prospective cohort study of two large cohorts.	2309 individuals from the Rotterdam and the LifeLines-DEEP cohort.	Fasting plasma metabolite profiling with 1H-nuclear magnetic resonance (NMR). Fecal sample gut microbiota profiling with 16s rRNA gene sequencing.	Significant ($p < 0.05$) associations between 18 families and genera of bacteria with VLDL of different sizes, 22 with HDL, 13 with HDL and VLDL, and 15 with serum triglycerides

Le Roy et al., 2019 [73]	Mouse study	Hypercholesterolemic female Apoe−/− and LDLr−/− mice	Depletion of all microbes in hypercholesterolemic mice with a combination of 4 antibiotics. Human feces intestinal microbiota transplantation. Further on mice: Plasma lipid and lipoprotein profile analysis. Bile acid synthesis measurement with labeled (14C) cholesterol dissolved in olive oil. Liver, ileum, and jejunum gene expression with qPCR. Sterol quantification in liver and bile. 16S rRNA gene sequencing of fecal gut microbiota.	Depletion of microbiota in mice raises cholesterol, mainly VLDL and LDL. Depletion also enhances liver uptake of cholesterol. Depletion increases cholesterol de novo synthesis by liver. Cholesterol level is transmissible in mice by microbiota transplantation from humans with altered cholesterol levels to microbiota depleted Apoe−/− mice.
Yun et al., 2020 [74]	Cross sectional study	1141 subjects from the Kangbuk Samsung Health Study in South Korea, divided into dyslipidemic (G0) and normal lipid (G1) groups based on total cholesterol, LDL-C, TG, HDL-C, ApoB and ApoA1 levels.	Anthropometric data. Blood sample after 10 h fasting of total cholesterol, triglycerides, LDL-C, HDL-C, and apoA1 (HDL particle component) and ApoB (LDL-C and other particle component) determination. Gut microbiota 16s rRNA gene sequencing of fecal sample.	The group with high TGs had lower alpha diversity indices (Shannon's index and Faith's phylogenetic diversity, both $p < 0.001$, Pielou's evenness $p < 0.030$). Abnormally low ApoA1 group had higher alpha diversity. No association with alpha diversity for other lipid parameters. 12 taxa associated with TGs: The high TGs group had high amount of the genus Fusobacterium and low levels of Oscillospira, which produces butyrate (a SCFA) 10 and 6 taxa were associated with ApoA1 and ApoB levels, respectively.

5.4 TARGETING DYSLIPIDEMIA AND DYSBIOSIS:

The abovementioned examples and other work have highlighted how alterations in gut microbiota can lead to maladies inclusive of dyslipidemia. This dysbiosis induced dyslipidemia and inflammation to cause cardiovascular diseases (CVDs). Changes in trimethylamine-n-oxide (TMAO) and butyrate due to dysbiosis have been outlined in atherogenesis by research (Lau, Srivatsav, Rizwan, Nashed, Liu, Shen and Akhtar, 2017).

Figure 8 below depicts the risk factors of CVD and gut dysbiosis (Lau, Srivatsav, Rizwan, Nashed, Liu, Shen and Akhtar, 2017: © 2017 by the authors. Licensee MDPI, Basel, Switzerland. *This article is an open access article distributed under the terms and conditions of the Creative Commons Attribution (CC BY) license: http://creativecommons.org/licenses/by/4.0/*):

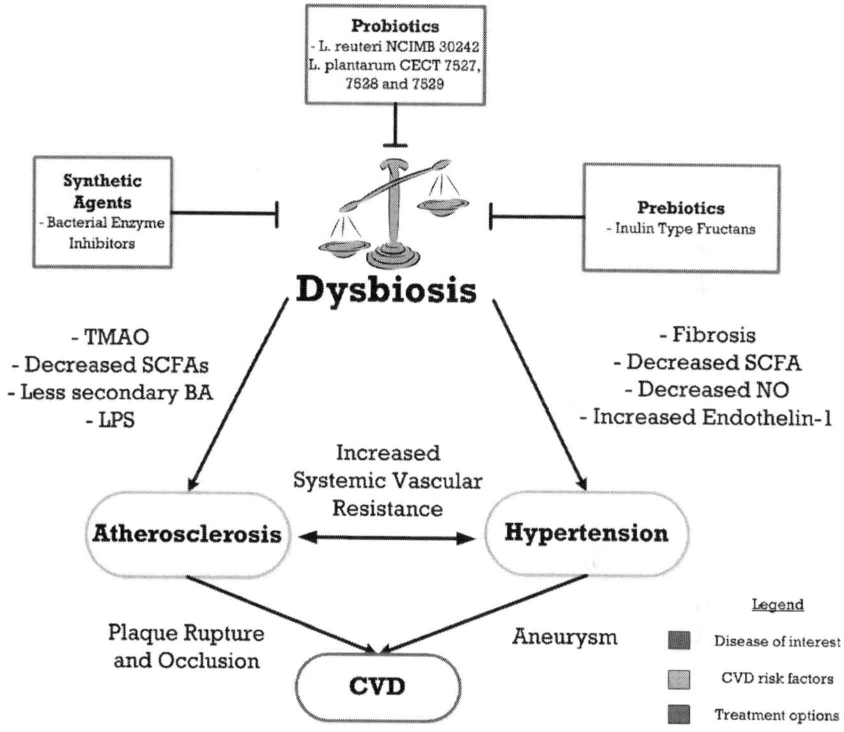

Health care research is exploring the use of ***probiotics*** or "friendly microbes" that maintain a healthy gut microbiota. Table 4 below summarizes research conducted on how probiotics have been employed to alter lipid levels in animal models and humans as reviewed and compiled by He and You (2020): *Key: CFU: colony-forming units; TC: total cholesterol; TG: triglyceride; HDL-C: high-density lipoprotein cholesterol; LDL-C: low-density lipoprotein cholesterol* (He and You: 2020: © 2020 Yan-Jun He and Chong-Ge You. *This is an open access article distributed under the Creative Commons Attribution License, which permits unrestricted use, distribution, and reproduction in any medium, provided the original work is properly cited*):

Author, year	Model (human/animal)	Gender	Probiotics	Dosage and duration	Changes in gut microbiota	Lipid changes
Falcinelli et al., 2015 [11]	Zebrafish	Female and male	*Lactobacillus rhamnosus*	1×10^6 CFU/ml 18 hours	Increased *Firmicutes*, *Lactobacillus*, and *Streptococcus thermophilus*; reduced *Actinomyces*	Reduced TC and TG
In Kim et al., 2019 [12]	C57BL/6 mice	Male	*Lactobacillus plantarum*LC27, *Bifidobacterium longum*LC67	1×10^9 CFU/ml 4 weeks	Reduced *Firmicutes* and *Proteobacteria*	Reduced TC and LDL-C; increased HDL-C
Wang et al., 2018 [21]	Human	No description	*Bifidobacterium bifidum* TMC3115	3×10^{10} CFU/mL 3 weeks	Increased *Firmicutes*; reduced *Bacteroides*	Reduced TC and LDL-C
Costabile et al., 2017 [22]	Human	34 females, 15 males	*Lactobacillus plantarum* ECGC 13110402	2×10^9 CFU/ml 12 weeks	No significant differences	Reduced TC and TG; increased HDL-C
Shin et al., 2018 [30]	Sprague Dawley rats	Male	A probiotic formulation containing 2 *Lactobacillus* strains, 3 *Bifidobacterium*, and *Streptococcus thermophilus* ST3	1×10^7 CFU/ml 8 weeks	Increased *Bacteroides*, *Lactobacillus*, and *Bifidobacterium*; reduced *Firmicutes*	Reduced TC, TG, and LDL-C
Salazar et al., 2019 [31]	C57BL/6 mice	Male	*B. animalis* IPLA R1 strain	5×10^8 CFU/ml 10 days	Increased *Bifidobacterium*	Reduced TG
wang et al., 2017 [32]	Chicks	Male	*Lactobacillus johnsonii*	1×10^6 CFU/ml 42 days	Increased *Bacteroides* and *Lactobacillus*; reduced *Escherichia coli*	Reduced TG and LDL-C increased HDL-C
Chen et al., 2015 [29]	Wistar rats	Male	*Lactobacillus rhamnosus* hsryfm 1301	10^9 CFU/ml 28 days	Increased *Bifidobacterium*, *Lactobacillus*, *Bacteroides*, and *Enterococcus*; reduced *Escherichia coli* and *Clostridium*	Reduced TC and TG
Park et al., 2018 [34]	C57BL/ 6JTacN mice	Male	*Lactobacillus rhamnosus* strain BFE5264	1×10^{10} CFU/ml 9 weeks	Increased *Mucispirillum*	Reduced TC and LDL-C; increased HDL-C

lyea et al, 2017 [35]	BALB/*c* mice	Male	*Lactobacillus fermentum* FTDC 8312	1×10^9 CFU/ml 7 weeks	Increased *Akkermansia, Desulfovibrio, Oscillospira*, and *Lactobacillus*	Reduced TC and LDL-C; increased HDL-C
Zhu et al., 2019 [38]	C57BL/6J mice	Male	*Enterococcus faecalis* ATCC19433	5×10^9 CFU/ml 4 weeks	Increased *Lactobacillus, Bifidobacterium, Akkermansia*, and *Bacteroides*; reduced *Firmicutes*	Reduced TC and LDL-C
Briand et al., 2019 [40]	Golden Syrian hamsters	Male	*Saccharomyces boulardii* CNCM I-745	3g/kg 39 days	Increased *Proteobacteria* and *Lentispharerae*; reduced *Firmicutes*, and *Tenericutes*	Reduced TC; reduced HDL-C at 21 days
Liu et al., 2017 [36]	Immature kunming rats	Male	*Lactic Acid Bacteria* 96	1×10^9 cfu/mL 5 weeks	Reduced *Escherichia coli*	Reduced TG, TC, and LDL-C
Ichim et al., 2016 [37]	C57BL/6J mice	No description	DBR:a proprietary blend of 9 probiotic organisms and 10 digestive enzymes	65 μg/ml 8 weeks	Increased *Lactobacilli*	Reduced LDL-C; increased HDL-C

Studies are unveiling the involvement of gut microbiota on cholesterol homeostasis. Apart from comprehending the mechanistic aspects, research can also cast light to design treatment regimens for aberrant cholesterol profiles by correcting gut dysbiosis. Latest advances like omics platforms can emerge as vital players for these purposes to look at personalized treatment employing our gut residents.

". . . our body is an ecosystem. This ecosystem must be maintained . . ." — Ilchi Lee, Belly Button Healing: Unlocking Your Second Brain for a Healthy Life

5.5 REFERENCES:

1. Lin, L., & Zhang, J. (2017). Role of intestinal microbiota and metabolites on gut homeostasis and human diseases. *BMC immunology*, *18*(1), 2. https://doi.org/10.1186/s12865-016-0187-3
2. Shivsanjeevi Sripathi and Prerna Pandey (2022) Understanding our Valuable Tenants: Gut Bacteria. Delve Publishing. ISBN: 9781774690444
3. Matthew WK Wong, Anbupalam Thalamuthu, Nady Braidy et al. Genetic and environmental determinants of variation in the plasma lipidome of older Australian twins. eLife 2020;9:e58954 DOI: 10.7554/eLife.58954
4. Kho ZY and Lal SK (2018) The Human Gut Microbiome – A Potential Controller of Wellness and Disease. *Front. Microbiol.* 9:1835. doi: 10.3389/fmicb.2018.01835
5. Bo, T, Shao, S, Wu, D, Niu, S, Zhao, J, Gao, L. Relative variations of gut microbiota in disordered cholesterol metabolism caused by high-cholesterol diet and host genetics. *MicrobiologyOpen*. 2017; 6:e491. https://doi.org/10.1002/mbo3.491
6. Le Roy, T., Lécuyer, E., Chassaing, B. et al. The intestinal microbiota regulates host cholesterol homeostasis. BMC Biol 17, 94 (2019). https://doi.org/10.1186/s12915-019-0715-8
7. Villette R, KC P, Beliard S, Salas Tapia MF, Rainteau D, Guerin M and Lesnik P (2020) Unraveling Host-Gut Microbiota Dialogue and Its Impact on Cholesterol Levels. *Front. Pharmacol.* 11:278. doi: 10.3389/fphar.2020.00278
8. de Moraes, A. C., Fernandes, G. R., da Silva, I. T., Almeida-Pititto, B., Gomes, E. P., Pereira, A. D., & Ferreira, S. R. (2017). Enterotype May Drive the Dietary-Associated Cardiometabolic Risk Factors. *Frontiers in cellular and infection microbiology*, *7*, 47. https://doi.org/10.3389/fcimb.2017.00047
9. Zwartjes, M., Gerdes, V., & Nieuwdorp, M. (2021). The Role of Gut Microbiota and Its Produced Metabolites in Obesity, Dyslipidemia, Adipocyte Dysfunction, and Its Interventions. *Metabolites*, *11*(8), 531. https://doi.org/10.3390/metabo11080531

10. Lau, K., Srivatsav, V., Rizwan, A., Nashed, A., Liu, R., Shen, R., & Akhtar, M. (2017). Bridging the Gap between Gut Microbial Dysbiosis and Cardiovascular Diseases. *Nutrients*, *9*(8), 859. https://doi.org/10.3390/nu9080859
11. He, Y. J., & You, C. G. (2020). The Potential Role of Gut Microbiota in the Prevention and Treatment of Lipid Metabolism Disorders. *International journal of endocrinology*, *2020*, 8601796. https://doi.org/10.1155/2020/8601796

CHAPTER 6

Cholesterol and the Reproductive System

(Prerna Pandey and ShivSanjeevi Sripathi)

Contents

6.1 INTRODUCTION

Cholesterol plays a significant role in human reproduction. Lipoprotein cholesterol is the major source of substrate for steroidogenesis, which is crucial for gonadotropin production. Cholesterol is a necessary precursor for the biosynthesis of steroid hormones in steroidogenic cells of the gonads, adrenal glands, and placenta, which, in turn regulate vitally important physiologic functions, such as controlling reproductive pathways, the formation of secondary sexual characteristics, and controlling salt balance. Cholesterol substrate is essential for the activated steroidogenic cells. A variety of mechanisms in the cell work together to ensure that the cholesterol pool is replenished and available in an adequate amount. As a result of such mechanisms, lipoprotein cholesterol is taken up by specific receptors (i.e., low-density lipoprotein receptors [LDLR] and/or scavenger receptor class B type I [SR-BI]), esters are hydrolyzed by hydrolases including hormone sensitive lipase, and cholesterol is synthesized by a multistep process starting with acetate, and the cholesterol is eventually stored in the cell (DeAngelis et al.,2014).

6.2 DISCOVERY OF STEROLS THAT AFFECT MEIOSIS *IN VITRO*

Several intermediates are involved in the cholesterol biosynthesis pathway. It has long been known that early intermediates of cholesterol synthesis serve as precursors of essential compounds such as dolichol, heme A, and coenzyme Q10. A complex single enzyme reaction led to the formation of the first steroidal intermediate, lanosterol, from squalene. Late sterol intermediates, also known as neutral sterols, were long believed to be used exclusively in cholesterol synthesis. However, in 1995, a potential functional role of neutral sterols in mammals was proposed. As a result of screening for naturally occurring meiosis-activating compounds, meiosis-activating sterols (MAS) have been identified. In vitro experiments have demonstrated that MAS triggers the resumption of meiosis in cultured mouse oocytes. Initial identification of two structurally related sterols with similar biological effects was primarily directed to human preovulatory ovarian follicular fluid (FF-MAS) and testis of adult bulls (T-MAS). The chemical names are 4,4-dimethyl-5α-cholest-8,14,24-triene-3β-ol for FF-MAS and 4,4-dimethyl-5α-cholest-8,24-diene-3β-ol for T-MAS. A chemically synthesized sterol, MAS-414, and its close relative, MAS-412, were able to induce meiosis in vitro in a manner similar to that of naturally occurring sterols. It was the first time sterol intermediates had direct biological roles in addition to merely serving as transitory intermediates in a chain of reactions leading to the production of cholesterol. In vitro assays have shown that MAS accumulate primarily in the gonadal tissue of both sexes and that they promote meiosis in a non-sex-specific manner. Based on these initial biochemical studies, MAS was suggested to play an important role in reproduction in vivo.

Numerous in vitro studies investigating the potential role of MAS in oocyte maturation followed the identification of MAS's meiosis-activating capacity. As a result of a surge in gonadotropins during ovulation, MAS produced by cumulus cells serves as a paracrine factor to stimulate oocyte meiosis. It was originally thought that MAS could be used to treat infertility and to develop new contraceptives without relying on sex steroids or similar compounds; however, later evidence revealed that MAS are not required to regrow meiosis, as mediated by gonadotropin hormones. In a context of meiosis-activating activities, eight endogenous MAS molecules with similar structures have since been identified (Grondahl, 2008), demonstrating the potency of FF-MAS and related MAS analogs in vitro. MAS compounds, however, have also been found to have a positive effect on cytoplasmic maturation, a crucial process during the final stages of embryonic development that facilitates nuclear *maturation, fertilization, and proper early embryonic development. Contrary to the originally proposed role of MAS in the resumption of meiosis, it has now been suggested that MAS may also play a role in meiotic nuclear processes and cytoplasmic maturation (Grondahl, 2008), however, the exact mechanism of MAS in female reproduction remains unclear.

6.3 ACCUMULATION OF MEIOSIS-ACTIVATING STEROLS IN THE TESTIS AND SPERM

The function of MAS in oocytes has been extensively studied, but studies investigating its function in spermatogenesis are rare. Experimental evidence demonstrated in vitro that a diffusible component contained in adult mouse testes triggered non-differentiated fetal male germ cells to enter meiosis. Although the role of MAS in male germ cell development was frequently suggested following its discovery in 1995, no causal relationship has been established in vivo or in vitro. A majority of studies have focused on quantifying MAS or cytochrome P450 14α-demethylase (CYP51), an enzyme responsible for producing MAS, in the male reproductive tract and correlating these parameters with reproductive function. According to the first reports of MAS accumulation in adult bull, mouse, and horse testes, MAS accumulates at concentrations exceeding 30μg/g (parts per million [ppm]); however, lower concentrations of MAS were detected in isolated rat seminiferous tubules (15 ppm). A 40-fold higher concentration of T-MAS was found in the rat testes than in the liver, indicating that T-MAS is the major sterol intermediate in the testis. T-MAS concentration in the rat testis increased eightfold from 19 to 70 days of age (i.e., 6.1 to 49ppm) during sexual maturation. According to another study, T-MAS increased in rat testes by fourfold during puberty. Similar changes are not observed in other sterol precursors during puberty, suggesting that T-MAS plays an important role in spermatogenesis.

Further evidence of accumulation of T-MAS was found in sexually maturing stallions. Based on a study of 16 stallions aged from 1 to 12 years, there was a positive correlation between testicular T-MAS concentration and testicular weight as well as a negative correlation between testicular T-MAS concentration and cryptorchidism after puberty. T-MAS concentrations in mature studs were seven times higher than those in immature males (19.3 ppm vs. 2.8 ppm). Human spermatozoa have yielded a 4-fold higher T-MAS concentration (2 ppm) than those from ejaculate (< 0.5 ppm), indicating that MAS is accumulated in sperm cells as opposed to seminal plasma. Nevertheless, measurements of MAS and several other sterol intermediates in guinea pig testes showed that, by quantity, T-MAS ranked third (15.6 ppm) after lathosterol (23.5 ppm) and desmosterol (21.4 ppm). There was a 15-fold increase in T-MAS in the testicle compared with the liver. In the adult mouse testis (10-12 weeks), T-MAS was found to be the most abundant sterol (71 ppm), followed by desmosterol (15 ppm). Testis contain at least 70 times more T-MAS than liver. Due to the fact that desmosterol was found to be higher than T-MAS (52 vs. 40 ppm) in these HPLC-based sterol profiling studies that used mice testes of the same age, these results differ from recent studies executed using gas chromatography-mass spectrometry. Nonetheless, TMAS has been found in mature testes and sperm from a wide range of mammals, providing a compelling argument for the potential importance of MAS to sperm development (Keber et al.,2013).

6.4 CHOLESTEROL AND FEMALE FERTILITY

6.4.1 Cholesterol Metabolism and Ovarian Function

The metabolism of lipoproteins from liver and small intestine to ovarian tissue has been extensively studied since Nobel Prize winners Michael Brown and Joseph Goldstein elucidated the mechanism of cellular uptake of cholesterol, suggesting the possibility of nutritional modulation of ovarian steroidogenesis. Using Watanabe heritable hyperlipidemic rabbits with mutated LDL receptors, Robins et al. revealed lower ovarian steroid production and lowered fertilization rates, indicating for the first time that ovarian steroidogenic processes are dependent on the LDL receptors. This was then confirmed by the fact that cholesterol intake via scavenger receptor class B type I (SR-BI) is essential for mouse fertility and progesterone secretion in human granulosa cells as well. Increasing evidence links obesity and metabolic syndrome to fertility problems in humans: impaired ovulation, oocyte quality, and embryonic development (Robker et al.,2011).

A fascinating sterol has been implicated in oocyte formation along with cholesterol as a precursor to steroid hormones. This sterol is called follicular fluid meiosis-activating sterol (FF-MAS, 4,4-dimethyl-5α-cholest-8,14,24-trien-3β-ol). Upon LH treatment, following FSH priming, its concentration increases in the ovary, stimulating the resumption of meiosis of the oocyte, maintained at the prophase of the first meiotic division. In view of the absence of FSH receptors on the oocyte, it has been suggested that FF-MAS may be a signal from the cumulus cells to the oocyte, thereby promoting meiosis resumption.

Overall, obese patients undergoing in vitro fertilization (IVF) have lower pregnancy rates (Maheshwari et al.,2007) and body mass index (BMI) is strongly associated with failure in IVF, suggesting the nutritional status, particularly lipids, can be a risk factor for ovulation induction.

6.4.2 Cholesterol and Uterine Physiology

In 1956 it was proposed that progesterone, a hormone synthesized by the body, may influence pregnancy until term. Steroids have been known to influence pregnancy since the 1920s, and the hypothesis of the "progesterone block" can be traced back to this period. Currently, it is well established that the progesterone-to-estrogen ratio is the main determinant of uterine contractions in most mammals. At term, the ratio of progesterone to estrogen rapidly decreases, resulting in uterine contractions, which is why progesterone concentrations during pregnancy are high. It is unclear, however, what molecular signal underlies the onset of parturition, which induces the fall of estrogen to progesterone ratios. A possible cause involves the production of surfactant protein-A (SP-A), which arises from mature fetal lung tissues and pro-inflammatory factors (Mendelson,2009).

Uterine physiology may be affected by lipid status. Having a high BMI certainly increases the chance of a cesarean delivery (CD). A meta-analysis showed odds ratios for CD increasing to 1.53 [95% confidence interval (CI): 1.48-1.58], 2.26 (95%CI: 2.04-2.51) and 3.38 (95%CI: 2.49-

4.57) for overweight (25<BMI<30), obese (30<BMI<35) and morbidly obese (BMI>35), respectively. The defect in contractile activity of the uterus increases the risk of CD in both overweight and obese women. The study found a correlation between BMI and myometrium contractility. Cholesterol metabolism is an interesting candidate among the molecular mechanisms of this association. In fact, during a normal pregnancy, circulating cholesterol concentrations increase until the due date. In contrast, a fetus with a non-controlled maternal hypercholesterolemia may be negatively affected, suggesting that, although essential for a pregnancy, maternal cholesterol concentrations must be carefully controlled during pregnancy (Lobaccaro et al.,2012).

There is an inverse correlation between the cholesterol content of uterine muscular cells and their contractility in animal models, and between spontaneous contractility of myometrium and BMI, according to ex vivo studies (Mouzat et al.,2012). On the other hand, cholesterol is critical to oxytocin receptor activity and stability, which play an important role in uterine contractions, which highlights the importance of maintaining a tight control of cholesterol metabolism within uterine myocytes as well.

6.5 THE ROLE OF CHOLESTEROL IN SERTOLI CELLS DURING SPERMATOGENESIS

Spermatogenesis relies on lipids and cholesterols to function as 'fuel' for Sertoli cells and to remodel the membrane of developing germ cells. A series of enzyme reactions is involved in the biosynthesis of cholesterol. Screening of naturally occurring compounds revealed that meiosis activating sterols had strong meiosis activating potency. In addition, ABCA1 and scavenger receptor class B member 1 (SR-BI) are responsible for regulating cholesterol content in Sertoli cells (Umehara et al.,2016).

6.5.1 The Transport of Cholesterol in Sertoli Cells

Sertoli cells provide most of the nutrients for spermatogenesis, such as lipids. Acetate can be used in vitro by Sertoli cells to synthesize cholesterol, despite this, in vivo, cholesterol from this source was not sufficient for steroidogenesis. Since spermatogenesis requires far more cholesterol than Sertoli cells are capable of synthesising, specialized cholesterol transporters such as SR-BI facilitate the uptake of cholesterol from blood streams. The findings of Fofana et al. (2000) suggest that HDL serves as the main source of cholesterol for Sertoli cells, which is further supported by the observation that the membrane separating seminiferous tubules from blood capillaries prevents the entry of LDL, but permits HDL entry (Fofana et al.,2000). Sertoli cells in rats receive cholesterol from HDL primarily through apolipoprotein E-dependent pathways. Additionally, Sertoli cells are capable of extracting cholesterol from phagocytosed apoptotic germ cells as well as remnants with a high lipid content. It is crucial to maintain cholesterol equilibrium that excess cholesterol is esterified and stored in lipid droplets to eliminate toxicity. In addition to their high mobility and dynamic nature, lipid droplets are also reported to interact with cell junctions and organelles via transient or stable surface exchanges, ensuring the proper synthesis, uptake, usage, and elimination of cholesterol (Pelletier,2011). In addition, Sertoli cells regulate cholesterol homeostasis via reverse cholesterol transport (RCT). Various procedures are utilized in RCT to enable the efflux of cholesterol from nonhepatic peripheral tissue to HDL and promote the transport of cholesterol from the nonhepatic peripheral tissues back to the liver via the plasma to maintain cholesterol homeostasis, which primarily involves ABCA1 cholesterol transporters. Cholesterol esters (CE) accumulate excessively in Sertoli cells, disrupting cholesterol homeostasis and ultimately resulting in almost complete infertility. The phenomenon has been observed in various knockout mouse models that lack functional nuclear receptors, such as RXR, and ABCA1. Due to ABCA1's established liver X receptor (LXR) target gene status, ABCA1 deficiency reduces cholesterol efflux from Sertoli cells and results in increased CE production in double knockout mice (LXR α/β/). Consequently, Sertoli cells regulate every stage of cholesterol metabolism, including cholesterol uptake, efflux, storage, and recycling, which suggests that Sertoli cells should be considered in the study of spermatogenesis. An illustration of cholesterol transport in Sertoli cells is shown in Figure 1.

6.5.2 The Role of Cholesterol Metabolism in Spermatogenesis

It is well known that cholesterol plays an important role in sperm production and male fertility. Spermatogenesis involves a series of phases involving differentiation and proliferation, which are further divided into spermatogenic, meiotic and mitotic stages. There are different types of cells at each stage, such as spermatids, spermatocytes and spermatogonia. Throughout spermatogenesis, increases in lipid droplets are observed, demonstrating a correlation between lipid metabolism alterations and fertility during spermatogenesis. The process of steroidogenesis also demands large amounts of cholesterol, while in the seminiferous tubules, cholesterol is required for the differentiation of germinal cells into spermatozooids (gametogenesis/spermatogenesis). Additionally, considerable evidence indicates that cholesterol is necessary for both males and females to develop fertility and gametes. A lack of 24dehydrocholesterol reductase in male mice, which encodes a cholesterol biosynthetic enzyme, results in infertility. Consequently, cholesterol is necessary to produce large numbers of germ cells during spermatogenesis. When sperm are released from the seminiferous epithelium, their plasma membranes are also loaded with cholesterol. An experiment that determined the role of cholesterol de novo synthesis previously determined that the incorporation rate of 14C acetate into cholesterol was increased during development of the Pachytene, Leptotene, and Zygotene stages. As a result, these germ cells produced more cholesterol de novo, which is associated with a larger diameter and surface area. The rate of cholesterol synthesis tends to decrease in late pachynema and remains low throughout the rest of spermatogenesis, including sperm maturation. Unlike cholesterol, the incorporation capacity of acetate into dolichol was enhanced in round spermatids and late pachytene spermatocytes, but then diminished and remained low in mature sperm cells. As a result of these observations, cholesterol synthesis in round spermatids and pachytene spermatocytes may occur before the synthesis of dolichol, which is an integral constituent of membrane glycoproteins. It has not been possible to determine the precise role of cholesterol metabolism in germ cells, partly because of the complex interactions between germ cells and supporting Sertoli cells.

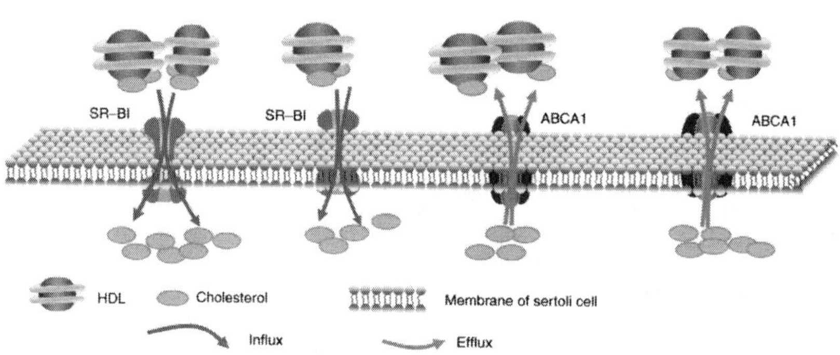

Figure 1. Cholesterol is transported by Sertoli cells. In Sertoli cells, cholesterol is transported primarily by ABCA1 and SR-BI, two important cholesterol transporters on the membrane. A large amount of cholesterol is absorbed by Sertoli cells through the uptake of HDL cholesterol esters. Moreover, ABCA1 controls cholesterol homeostasis by effluxing surplus cholesterol from Sertoli cells. It is composed of the ATPbinding cassette subfamily A member 1, SR-BI, scavenger receptor class B member I, and HDL, high-density lipoprotein. (Shi, J. F., Li, Y. K., Ren, K., Xie, Y. J., Yin, W. D., & Mo, Z. C. (2018). Characterization of cholesterol metabolism in Sertoli cells and spermatogenesis (Review). *Molecular medicine reports*, *17*(1), 705–713.)

6.6 SYNTHESIS OF CHOLESTEROL IN GERM CELLS DURING SPERMATOGENESIS

During spermatogenesis, cholesterol is required for the extensive production of germ cells. Cholesterol is thought to be highly concentrated in sperm plasma membranes after release from the seminiferous epithelium. The early experiments that examined de novo cholesterol synthesis revealed an increased incorporation of 14C acetate into cholesterol during the leptotene, zygotene, and pachytene development stages for spermatocytes. This indicates that de novo cholesterol synthesis is occurring in these germ cells, which is in line with the increase in the diameter and size of the germ cells during development, which results in the synthesis of new plasma membranes. In late pachynema, the capacity to synthesize cholesterol decreased and remained low throughout the rest of spermatogenesis and mature sperm. Spermatocytes do not grow significantly after the pachytene stage, but instead differentiate into diplotene spermatocytes, which undergo meiotic divisions to form round spermatids. Acetate incorporation rate into dolichol was higher in pachytene spermatocytes and around spermatids than cholesterol, but decreased and remained low in mature sperm. It is likely that the early stages of cholesterol synthesis in pachytene spermatocytes and round spermatoids precede the synthesis of dolichol, a crucial component in protein glycosylation and the production of specific membrane glycoproteins. Despite this, it has been difficult to determine precisely the role of de novo cholesterol synthesis in germ cells, mainly because the relationship between the germ cells and supporting Sertoli cells is complex (Bielska et al.,2014).

6.7 FUNCTIONAL ROLE OF STEROLS IN SPERM DEVELOPMENT

As spermatozoa develop, membrane structure undergoes extensive remodeling, altering permeability and fluidity of the membrane at different stages. A sperm cell must maintain integrity and acquire a high level of motility while maturing. The membrane structure of the female reproductive tract undergoes a dramatic change prior to the fusion of the oocyte with its acrosomal content, which allows the extrusion of the acrosomal content. At distinct stages of the maturation process, sperm membrane lipid content is precisely controlled in order to acquire highly specialized membrane function. Unique molecular events that resulted in stage-specific alterations of sperm membrane sterol content clearly demonstrate the functional importance of sterol composition during sperm maturation.

It is known that high levels of different sterol intermediates accumulate in the testis in addition to cholesterol. Despite not knowing their exact function, they are believed to play an important role in the modification of membrane properties. It is not understood why cholesterol is selected as the major sterol in most animal cells; however, it is generally accepted that cholesterol is the most effective sterol at modulating membrane properties. Due to its structure, cholesterol can be tightly bundled with the tails of neighboring phospholipids. By means of its rigid ring system and its ability to fill interstitial spaces, this can increase order within liquid membranes. As a result of higher lipid packing order, a functional barrier of permeability is formed in the membrane. Cholesterol concentration is generally highest in immature and immotile sperm collected from the caput epididymis or testis. Currently, cholesterol is believed to function as a membrane stabilizer that prevents premature release of proteolytic enzymes from the acrosome. During sperm maturation, cholesterol content decreased in most species, while the content of other sterols, such as desmosterol, increased. Sperm function is critically dependent on cholesterol, which has been demonstrated in patients with idiopathic infertility who display a higher cholesterol to phospholipid ratio.

In humans, high cholesterol levels in sperm negatively correlate with sperm quality (Zalata et al.,2010). Furthermore, cholesterol is believed to play a specific role in the sperm raft. During capacitation, there may be a tendency for cholesterol to efflux preferentially from the nonraft pool to stimulate the formation of larger rafts over the anterior sperm head.

6.8 IMPORTANCE OF CHOLESTEROL DURING SPERM MATURATION IN THE EPIDIDYMIS

Upon sperm release from the seminiferous epithelium, it is immotile and unable to fertilize an oocyte. The spermatozoa acquire their full fertilizing capacity during their transit through the epididymis and in the female reproductive tract through capacitation. Capacitation involves the activation of bicarbonate-dependent adenylyl cyclase, which results in the elevation of cAMP and tyrosine phosphorylation of a subset of flagellar proteins, which then leads to changes in sperm motility and acrosomal responsivenes. It is highly dependent on changes in plasma membrane cholesterol for these events to occur. Spermatozoa formed in the testis travel through the caput epididymis, reach the corpus, and then fall into the caudal region.

In order to fertilize, sperm cells must be collected from the cauda epididymis of different species. The cholesterol excreted by the epididymal epithelium contributes to the maturation of transiting sperm, according to an accepted theory (Saez et al.,2011). Additionally, the epididymis is capable of synthesizing cholesterol from scratch. In the mouse epididymis, cholesterol trafficking is regulated by LXR, as it is in Sertoli cells. In the lumen, cholesterol is likely to be effluxed by ABCA1 and ABCG1, which are well known LXR targets. The epididymal epithelial cells secrete a small membrane vesicle called epididymosomes (Fig. 2), which are mainly made up of sphingomyelin and various amounts of cholesterol. An important role for dietary cholesterol in epididymal sperm maturation has been suggested recently (Ouvrier et al.,2011). Dietary cholesterol overload in a LXR-deficient mouse model led to complete infertility due to lipid droplets accumulating in the epididymal duct smooth muscle cells. Hematological examination of the sperm, testicular weight, and sperm count showed that Spermatogenesis was not impacted. The high cholesterol diet reduced testosterone levels significantly in double knockout *Nr1h2* $^{-/-}$;*Nr1h3* $^{-/-}$ animals as well (Ouvrier et al.,2011). There is no direct evidence that dietary cholesterol intake affects the endogenous cholesterol concentration in sperm plasma membranes of wild-type animals, as there was for dietary fatty acids.

Human epididymis-specific proteins (HE1-HE6) may regulate sperm cholesterol content as it travels through the epididymal tract. HE1 or Niemman-Pick C2, a secreted sterol-binding glycoprotein involved in sterol transport, is probably the best studied member of this group. At least some species have suggested that the HE1 homolog plays an important role in regulating cholesterol content in sperm membranes. Several species exhibit a reduction in cholesterol content in sperm membranes of about 50% during epididymal transit. The cholesterol content of some species increased during sperm maturation or remained unchanged. As demonstrated in human spermatozoa, the loss of cholesterol increases the fluidity of sperm membranes, which might be necessary for the final steps of sperm maturation in the female reproductive tract. Furthermore, the presence of prostatic secretory granules or prostasomes, vesicular formations which are secreted by the prostate, demonstrates the importance of stage-specific sterols in sperm membranes. Sperm plasma membranes and prostatesomes have different lipid compositions. They are particularly rich in sterols and sphingomyelin. Although their physiological function has been extensively debated by many researchers. The prostatesomes inside sperm can fuse with the plasma membrane (Fig. 2), where they are thought to increase motility and prevent early acrosome reactions (Burden et al.,2006). There are several mechanisms that control cholesterol content in sperm membranes during epididymal transit, indicating the importance of cholesterol for sperm function.

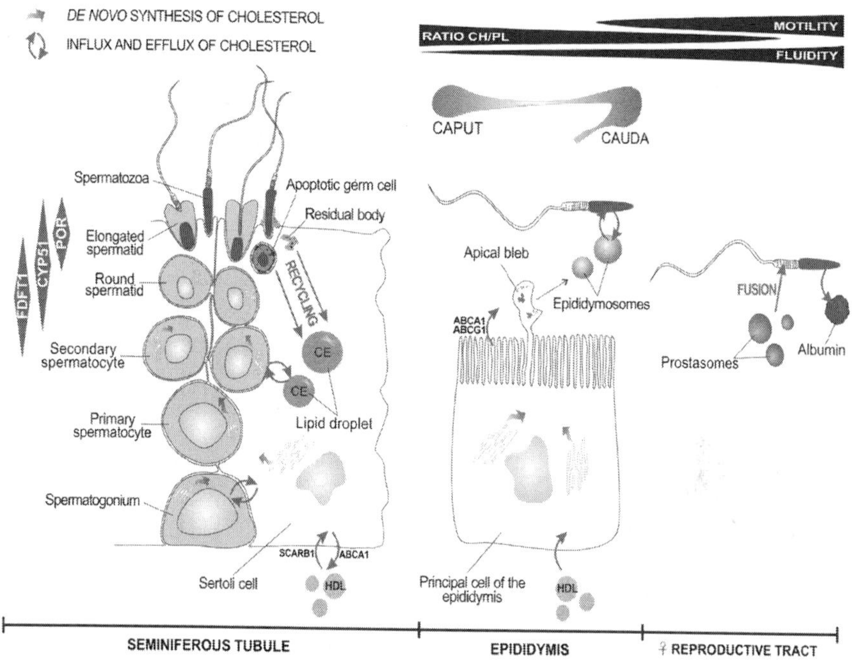

Figure 2: A schematic representation of the major sites of cholesterol synthesis during sperm maturation. During spermatogenesis, spermatocytes synthesize cholesterol de novo in a limited quantity required to synthesize new plasma membranes in seminiferous tubules. During spermatogenesis, gray boxes indicate the stages at which the cholesterogenic enzymes farnesyl diphosphate farnesyl transferase 1 (FDFT1), CYP51, and POR are expressed (left). Spermatogenesis is supported by Sertoli cells that provide cholesterol. Specifically, the Sertoli cells acquire cholesterol by synthesizing it from acetate or importing external cholesterol from HDL (Fofana et al., 2000) via specialized cholesterol transporters. Apoptotic germ cells and lipid-rich residual bodies may also be sources of cholesterol in Sertoli cells (Nakanishi and Shiratsuchi,2004).

It is possible to store cholesterol in lipid droplets by esterifying the excess cholesterol into cholesterol esters (CE) (Pelletier,2011). Cholesterol can be efluxed to HDL by reverse cholesterol transport. Spermatozoa in the testis migrate to the caput epididymis and then to the caudal region. Epididymis synthesizes cholesterol de novo, or it can import cholesterol from the circulation (Ouvrier et al.,2011). ABCA1 and ATP-binding cassette subfamily G member 1 (ABCG1) may efflux cholesterol into the epididymal lumen. There are small membranous vesicles found in the epididymis known as epididymosomes that may serve as a source of cholesterol exchange with maturing sperm cells. The ratio of cholesterol (CH) to phospholipids (PL) in sperm membranes decreases during epididymal transit in several species. Sperm motility and fluidity increase after cholesterol is lost from the sperm membrane. In the female reproductive tract, cholesterol content is further decreased by a process known as capacitation.

Capacitation occurs when cholesterol is taken up by albumin. The prostasomes in ejaculate are capable of fusing with the membranes of spermatozoa, increasing motility, and preventing early acrosome reactions (Burden et al.,2006). Several mammalian species were studied in vitro and in vivo to compile this scheme (Keber R, Rozman D, Horvat S. Sterols in spermatogenesis and sperm maturation. J Lipid Res. 2013 Jan;54(1):20-33.)

6.9 DYNAMICS OF SPERM MEMBRANE CHOLESTEROL IN THE FEMALE REPRODUCTIVE TRACT

The ejaculated sperm are motile, but they are not able to fertilize the egg. In the female reproductive tract, the process of capacitation activates Sperm. During capacitation, the sperm head undergoes remodeling, preparing it to bind to the zona pellucida (ZP) and undergo the ZP-induced acrosomal reaction. The first step to enabling capacitation is an organized destabilization of the plasma membrane, which also involves the redistribution of cholesterol, lipids, and proteins. In the apical ridge area of the sperm head plasma membrane, ZP-binding proteins cluster together during capacitation. Loss of cholesterol from the plasma membrane of the sperm is a necessary event during capacitation (Travis and Kopf,2002). The efflux of cholesterol is facilitated by albumin (Fig. 2) and high bicarbonate levels, which facilitate lipoprotein-mediated cholesterol efflux. It was found that cholesterol was removed only from the nonraft membrane portion of the sperm plasma membrane, while cholesterol content in the raft membrane fraction remained unchanged. Depletion of cholesterol at the sperm surface may be important for remodeling lipid rafts (Kawano et al.,2011). In parallel with cholesterol efflux, phospholipids were inserted into the sperm membrane, decreasing the cholesterol to phospholipid ratio (Fig. 2). Like sterols, the fatty acid composition of sperm during maturation is also highly modified. Combining these events increases membrane fluidity and facilitates Ca2+ influx into sperm, triggering the acrosomal reaction and oocyte fusion.

Different species have significantly different cholesterol contents in their sperm plasma membranes. Human sperm contain varying amounts of cholesterol, even within the same ejaculate. There is evidence, however, that cholesterol levels in sperm membranes are linked to sperm morphology and fertility potential. A mouse knockout model demonstrating the role of cholesterol in reproduction was also created by the complete ablation of the 24-dehydrocholesterol reductase gene (Dhcr24), which converts desmosterol to cholesterol. Only trace amounts of cholesterol were found in these animals, and they were unable to reproduce.

6.10 ROLE OF CHOLESTEROL EFFLUX IN SIGNAL TRANSDUCTION LEADING TO CAPACITATION

Several groups hypothesized that the requirement for serum albumin in defined media for capacitation was because albumin serves as a sink for cholesterol removal from the sperm plasma membrane. It is likely that the removal of this sterol accounts for the changes in membrane fluidity observed during capacitation and the subsequent reduction in the membrane cholesterol/phospholipid ratio. Cellular function is likely to be affected by such changes in membrane dynamics. Based on the fact that serum albumin reduces sperm membrane sterol levels following albumin exposure, this suggests that serum albumin mediates cholesterol efflux by acting as a cholesterol acceptor. Interestingly, the action of serum albumin, HDL, and β-cyclodextrins as cholesterol acceptors is somehow coupled to the cAMP dependent pathway described below (Figure 3).

What role might sterol content play in regulating transmembrane signaling resulting in capacitation? It has been demonstrated in biophysical studies that cholesterol alters the bulk properties of biological membranes. This sterol, for example, can increase the orientation order of membrane lipid hydrocarbon chains, making it harder for membrane proteins to undergo conformational changes by rendering the surrounding membrane less fluid (Travis and Kopf,2002). Cholesterol concentrations above a certain level can interfere with capacitation indirectly by diminishing conformational freedom, which inhibits sperm surface protein activity. It is also possible that cholesterol directly affects transmembrane signaling proteins. Similarly, ion transporters and effector enzymes like sAC may be affected by either or both of these effects of cholesterol, as illustrated in Figure 3. Despite the complexity of this story, it is likely to involve additional lipid and sterol species, especially given evidence that phospholipids move across distinct sperm membrane subdomains during capacitation.

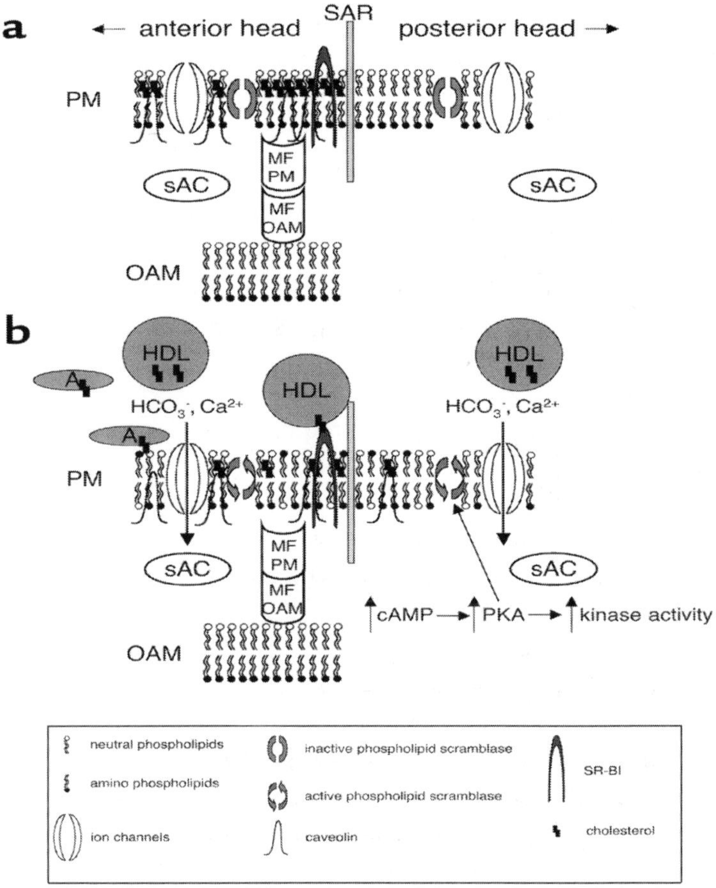

Figure 3: An overview of the structure and function of cholesterol in the membranes of the sperm head. A schematic diagram showing sperm incubated under noncapacitating conditions (a) and under capacitating conditions (b). In this model, membrane rafts (regions enriched in sterols) are arranged in the anterior head's plasma membrane. Fusion proteins attached to membranes are inactive due to the caveolin tether. Membrane leaflet phospholipid distribution is asymmetrical because scramblases do not function. Cholesterol is removed from the body via a specific pathway (HDL - SR-BI) as well as a nonspecific one (albumin) under capacitating conditions (bicarbonate, calcium, albumin, HDLs). The activation of sAC elevates cAMP levels, PKA activity, and stimulates downstream kinases. As a result of this signaling, phospholipid scrambling increases, causing disordered distribution of amino- and neutral phospholipids. Combined with the increase in membrane fluidity caused by sterol efflux, this change results in lateral movement of some sterols and caveolin from the anterior to the posterior head.

The loss of sterols in the membrane disrupts the interaction between caveolin and the membrane fusion proteins, leading to their activation. Within the subacrosomal ring (SAR), the plasma membrane (PM) and outer acrosomal membrane (OAM) are shown. The anion and the cation are shown as passing through the same ion channel in b. In the membrane, cholesterol can exist free or can be associated with cholesterol acceptors such as HDLs or albumin (A). This picture shows the cytoplasm of sperm containing sAC, a novel HCO3-activated adenylyl cyclase. Both the PM and the OAM can associate with membrane fusion proteins (MF). (Travis, A. J., and G. S. Kopf . 2002. The role of cholesterol efflux in regulating the fertilization potential of mammalian spermatozoa. *J. Clin. Invest.* **110:** 731 – 736.)

6.11 THE GERM CELL LINEAGE

Spermatogenic differentiation of male diploid germ cells into haploid gametes is a highly coordinated process. During spermatogenesis, type A spermatogonia are differentiated in contact with the basement membrane of the seminiferous tubules during the proliferative phase. As (a single) spermatogonia can replenish the population of undifferentiated germ cells. Spermatogonia can also participate in the differentiation process. During meiosis, spermatogonia divide into diploid and haploid cells, called secondary spermatocytes or spermatocyte II, which eventually develop into round spermatoids. Spermiogenesis is the final step of spermatogenesis after meiosis. There will be chromatin compaction in the spermatid nucleus due to reorganization. First of all, histones are replaced by small basic transition proteins (TNP1 and TNP2). To protect the genome from physical and chemical attacks, transition proteins will be replaced by specific testis nuclear proteins, protamines (PRM1 and PRM2).

During spermatogenesis, cholesterol metabolism and fertility are intimately linked. The mass production of germ cells requires cholesterol during spermatogenesis. During the development of zygotene, pachytene, and leptotene, cholesterol de novo synthesis was increased, which was associated with increased diameter and surface area of germ cells. As spermatogenesis progresses, cholesterol synthesis tends to decrease and remain low. In the last stages of spermatogenesis, the hormone-sensitive lipase (HSL) hydrolyzes cholesterol esters. HSL has been found in elongating spermatids and spermatozoa. Among Hsl−/− mice, the most notable phenotypic difference is oligospermia causing male sterility. Hsl−/− mice accumulate diacylglycerols and cholesterol esters in their testes. Scientists have demonstrated that spermatogenesis stops just before spermiogenesis elongation (Wang et al.,2017). The cells of Hsl−/− testes must be deprived of an important lipid metabolism pathway, causing them to differentiate abnormally. spermatozoa have a membrane made up of lipid bilayers, just like animal cells. A unique characteristic of the spermatic plasma membrane is the proportion of its constituents.

The spermatic membrane contains polyunsaturated fatty acids (PUFAs), which are necessary for the viability and mobility of the spermatozoon. Furthermore, PUFAs appear to play an important role in the process of membrane fusion between spermatozoon and oocyte. PUFAs contribute to the fluidity and flexibility of membranes. Docosahexaenoic acid (DHA), which accounts for a large portion of PUFAs in human cells, is crucial for fertility. In fact, scientists found that spermatozoa of asthenozoospermic men contained fewer polyunsaturated fatty acids (PUFAs) and more saturated fatty acids (SFAs) than sperm of normozoospermic men. There was also a negative correlation between body mass index and DHA, as well as a positive correlation between DHA and normal sperm parameters in another study (Andersen et al.,2016). Plasma membranes contain sterols as the second most important component. Many species of animals have spermatic membranes which contain cholesterol as their most abundant sterol. The seminiferous tubules of Spermatozoa contain a high amount of cholesterol. In fact, the germ cells can synthesize cholesterol de novo in the spermatocyte stage in order to increase the surface area of their membranes.

The sertoli cells also play a role in this process by providing the sperm cells with de novo synthesized cholesterol from the bloodstream. Recently, several species have been reported to contain oxidized cholesterol derivatives, including mice (Boerke et al.,2013). The fluidity of membranes could be affected by these oxysterols (Bielska et al.,2014). Besides phospholipids (PL), spermatic membrane contains sphingolipids as well. T During the process of capacitation in the female genital tract as well as during membrane fusion during fertilization, the lipid composition of spermatozoa is of great importance.

The lipid composition of the spermatic membrane is not definitive after spermiation, and it will change during the epididymal transit and capacitation process. Spermatozoa mature post-testicularly largely due to lipids. The lipid composition of the gamete is significantly rearranged during epididymal transit (Whitfield et al.,2015). In order to improve membrane fluidity, these modifications influence membrane dynamics. Rearrangements of these lipids are highly regulated. Spermatozoa receive proteins from the epididymis through lipid vesicles. Last but not least, capacitation is a phase of post-testicular maturation influenced by spermatic membrane lipid composition (Jin and Yang,2017). In order to maintain male fertility, lipid profiles of gametes must be regulated. According to a recent study, spermatozoa differ in lipid composition depending on the dietary intake of lipids (Jafaroghli et al.,2014), demonstrating a direct correlation between diet and gamete composition.

The purpose of mammalian spermatozoa is to fertilize a female oocyte. Spermatozoa are specialized and polarized cells. During the multistep process of post-testicular maturation, after spermiation, the fertilizing capacity is acquired. The maturation of Spermatozoa begins during their epididymal transit and ends in the genital tract. Despite its large size, the epididymis remains primarily responsible for maturing spermatozoa. It is necessary for embryonic development as well as for the acquisition of fertilizing power during epididymal maturation. The epididymal maturation process includes morphological, functional, and biochemical changes.

Signaling pathways involved in spermatogenesis include the LXR signalling pathways (Volle et al.,2007). Spermatogenesis occurs by balancing cell proliferation, cell differentiation, and cell death. Both proliferation and apoptosis are altered in Lxrβ−/− and Lxrα−/− mice, whereas these mice did not exhibit issues with fertility. When the proliferation rate of Lxrβ−/−mice was low, the apoptosis rate in Lxrα−/− mice was elevated, compensating for the lower proliferation rate. In the absence of both LXRs, proliferation dropped dramatically, and apoptosis increased. These mice may have lost all of their germ cells, explaining their infertility.

Researchers have recently investigated the role of LXRs in the physiopathology of non-obstructive azoospermia (NOA) patients' defective spermatogenesis (Rondanino et al.,2014). The abnormalities observed in NOA patients are similar to those found in Lxr−/− mice. The expression levels of Lxr mRNA in NOA specimens have been found to decrease in correlation with germ cell numbers. The authors found that Idol (Inducible Degrader of Insulin-Dependent LDL Receptor) and Srebp1c (Sterol regulatory element-binding protein 1), which are LXR target genes, decreased significantly.

Furthermore, a recent study showed that the G-coupled bile acids receptor TGR5 was expressed within the testis. Adult male fertility disorders are related to the BA-TGR5 signaling pathways that alter testicular epithelium integrity. It was found that after 4 months of dietary cholic acid (CA) supplementation, 25% of male mice were sterile and that 20% fewer pups were born per litter in fertile mice. It can be explained by the lower spermatozoa at the tail of the epididymis in men exposed to BAs. Interestingly, authors showed that Tgr5−/− mice were not affected by CA's deleterious effects on the testis. As a result of BAs exposure, the cell-cell interaction network is repressed via down-regulation of N-cadherin and connexin 43 expression. The result is sloughing of germ cells, rupture of the BTB, and spermatid apoptosis. Human studies have shown a reduction in expression of CX43 is associated with spermatogenesis defects (Matsuo et al.,2007).

A recent study revealed that FXRα controls the physiology of germ cells via the bile acid nuclear receptor (Martinot et al.,2017). The authors demonstrated that FXRα plays a key role in the establishment and maintenance of the undifferentiated germ cell pool, which in turn affects male fertility. The Fxrα−/− male shows an extended fertility with age. Fxrα−/− males maintain fertility capacity through a higher number of undifferentiated germcells during aging, resulting in a high spermatozoa count. FXRα is responsible for regulating the expression of several pluripotency factors in germ cells. Studies in vitro have shown that FXRα regulates Lin28 expression in germ cells.

6.12 OBESITY AND MALE FERTILITY

Infertility has become a global public health issue, affecting 15% of all couples. 70 million couples worldwide suffer from infertility or subfertility. Infertility is caused by male factors in 25% of cases. There is a lack of understanding of the etiology of low quality semen. Numerous physiological, environmental, and genetic factors play a role (Jungwirth et al.,2012).

The metabolic syndrome (MetS) consists of several risk factors, such as high blood pressure, high sugar levels, and high cholesterol levels. It is well established and well reviewed that MetS is associated with male hypogonadism (Dimopoulou et al.,2018). There is a low testosterone level in 20 to 64% of obese men according to cross-sectional studies. There is also evidence that low testosterone can increase the risk of diabetes mellitus and metabolic syndrome in men. In males, testosterone supplementation improves comorbidities such as obesity, MetS, and cardiovascular disease by improving inflammation, muscle mass, lipid profile, and sexual function (Dimopoulou et al.,2018).

Male obesity and components of the diet may have an important role in deregulating spermatogenesis, sperm maturation, or fertilizing ability in humans, as shown by human studies. It is possible for overweight or obese men to have decreased sperm quantity, quality, and motility (Sermondade et al.,2013). Researchers found that obese and lean patients had different acrosome reactions (Samavat et al.,2014). As a consequence of increased circulation levels of estradiol and sperm cholesterol, sperm acrosome reaction is impaired in obese men. Additionally, obese men may present a higher risk of sperm DNA damage and a lower rate of embryo implantation. In most human studies, obesity is only studied in relation to male fertility, not in relation to the lipid profile of the blood. Approximately 65% of infertile men exhibit hypercholesterolemia and/or triglyceridemia. Male fertility is still debated in relation to dyslipidemia. Despite the fact that cholesterol and lipid homeostasis are essential for male fertility, a recent study reported that hypercholesterolemia did not affect sperm concentration or motility in men (Eisenberg et al.,2015). Human sperm concentration does not correlate with serum total cholesterol or triglycerides (Lu et al.,2016).

A high cholesterol diet has been shown to impair fertility and lower sperm quality in some animal studies. Using a high fat diet (HFD) in rabbits is a good experimental model for the metabolic syndrome (MetS) (Maneschi et al.,2013). There are several diseases associated with this model including hyperglycemia, hypercholesterolemia, hypertension, hypogonadotropic hypogonadism, and penile alterations (Vignozzi et al.,2012). This model is very useful for studying the detrimental impact of MetS symptoms such as hypercholesterolemia on male fertility. A recent study demonstrated that supplementing testosterone with the FXRα agonist obeticholic acid (Vignozzi et al.,2012) normalized several MetS symptoms, including HFD-induced penile alterations. Sperm were abnormal in these animals (Fan et al.,2015). Tamoxifen, which is used in the treatment of idiopathic male infertility in humans, is shown to partially restore sperm motility but further reduce morphology and increase spontaneous acrosome reactions in HFD rabbits (Marchiani et al.,2015). Hypercholesterolemic rabbits are also found to have lower testicular efficiency due to a decrease in spermatogonial cells as well as an increase in germ cell apoptosis (Simon et al.,2017). These animals also suffered from spermiogenesis, the last step in spermatogenesis. An altered distribution of actin–alpha-tubulin–GM1 sphingolipid in post-meiotic germ cells from hypercholesterolemic rabbits was associated with abnormal acrosome, nucleus, and improper tail implantation. Recent studies have shown that a cholesterol-enriched diet increases lipid deposition in the seminiferous tubules and disrupts the BTB in rabbits (Morgan et al.,2014). This markedly reduced the total protein levels of the tight-junction protein 1 (ZO-1) and occludin as well as the distribution patterns of these proteins.

The High Fat Diet (HFD) also disrupted BTB in mice (Fan et al.,2015). According to studies conducted on male rodents exposed to HFD, motile spermatozoa are decreased, resulting in a decline in fertilization rates. Sperm motility decreased in the HFD group as compared to controls due to elevated levels of reactive oxygen species (ROS) and DNA damage. Researchers have found that HFD decreases testosterone production by decreasing the level of steroidogenic enzymes (Campos-Silva et al.,2017). Recent research suggests that HFD-induced fertility defects may be due to an increase in autophagy (Mu et al., 2017). Chloroquine inhibits autophagy and thus improves the decreased fertility of HFD male mice. It was also noted that obese, sub-fertile male patients exhibited excessive autophagy in their sperm samples.

There has also been some interest in the impact of paternal or maternal exposure to HFD on male offspring (Fullston et al., 2015). Despite their control diet consumption, exposed fathers to HFD preconception impair the motility of spermatozoa in their offspring (Fullston et al., 2015). Through pregnancy and lactation, feeding females a HFD causes male offspring to have lower levels of LH and testosterone (Rodríguez et al., 2015). Studies suggest that a calorie dense diet may cause reproductive defects to be amplified over generations if applicable to human health. A decade ago, it was suggested that derivatives of bile acids (BAs) could be used to treat metabolic diseases such as diabetes or obesity. A study has investigated the effects of long-term exposure to molecules that activate BA signaling pathways on male testicular physiology (Baptissart et al., 2016). In order to accomplish this, mice were fed HFD to induce obesity and treated concomitantly with BAs (Vega et al., 2015). The co-exposure of the HFD and BAs leads to a higher level of infertility, even if BAs improves some of the abnormalities caused by the HFD such as body weight, glycemia, and lipid profiles. This was accompanied by altered proliferation of germ cells, defective endocrine function, and abnormal interactions among cells within the seminiferous tubule.

6.13 CONCLUSION

It has long been suspected that cholesterol plays a role in female reproductive physiology, but its molecular basis was unknown. Cholesterol is an essential component of ovarian steroid biosynthesis and plays a role in fertility. The uterus relies on cholesterol to achieve correct contractions at term, but an excessive uterine cholesterol concentration has been associated with contractility defects (Lobaccaro et al.,2012).

Mammalian spermatogenesis involves the transformation of a diploid progenitor cell into a highly specialized spermatozoon. During spermatogenesis, the lipid composition of the membrane changes dynamically. Male and nongonadal tissues differ greatly in their cholesterol content as well as their intermediates. Testes and spermatozoa from several mammalian species accumulate cholesterol precursors such as testis meiosis-activating sterol and desmosterol. Moreover, cholesterogenic genes, particularly cytochrome P450 lanosterol 14α-demethylase, a meiosis-activating sterol-producing enzyme, show stage-specific expression patterns during spermatogenesis. Molecular differences in gene expression suggest that sterol compounds are regulated by cell type and temporal factors during spermatogenesis, which also involves dynamic interactions between germ and Sertoli cells. Further evidence of the importance of sterol compounds in sperm production is provided by the modulation of sterol composition in spermatozoal membranes during epididymal transit and in the female reproductive tract, both of which are essential for successful fertilization (Keber et al.,2013).

In order for testicular functions like steroidogenesis or Sertoli cell function to function correctly, cholesterol homeostasis is required. The altered concentration of plasma cholesterol can therefore affect male fertility and lead to infertility. Hyperlipidemia and metabolic syndrome are clearly linked to male fertility disorders. Some reproductive symptoms might be ameliorated by lipid-lowering medications. Epidemiological data, however, are scarce and controversial at the moment (Sedes et al.,2018).

The data collected here could aid in the development of new molecular diagnostic tests for infertility or determine whether the results obtained on the incriminated nuclear receptors LXRs and FXRα, which can be modulated by ligands, can be translated into strategies either to treat infertile men or to develop new methods for male contraception. In order to achieve such results, it will be necessary to develop testicular-specific LXR or FXR modulators, which can treat testicular disorders without causing associated metabolic disorders.

REFERENCES

1. Andersen JM, Rønning PO, Herning H, Bekken SD, Haugen TB, Witczak O. Fatty acid composition of spermatozoa is associated with BMI and with semen quality. Andrology (2016) 4:857–65.
2. Baptissart M,Martinot E, Vega A, Sédes L, Rouaisnel B, de Haze A, et al. Bile acid-FXRa pathways regulate male sexual maturation in mice. Oncotarget (2016) 7:19468–82.
3. Bielska AA, Olsen BN, Gale SE, Mydock-McGrane L, Krishnan K, Baker NA, et al. Side-chain oxysterols modulate cholesterol accessibility through membrane remodeling. Biochemistry (2014) 53:3042–51.
4. Boerke A, Brouwers JF, Olkkonen VM, van de Lest CHA, Sostaric E, Schoevers EJ, et al. Involvement of bicarbonate-induced radical signaling in oxysterol formation and sterol depletion of capacitating mammalian sperm during in vitro fertilization. Biol Reprod. (2013) 88:21.
5. Brown MS, Goldstein JL. A receptor-mediated pathway for cholesterol homeostasis. Science 1986, 232: 34-47.
6. Burden, H. P., C. H. Holmes, R. Persad , and K. Whittington . 2006. Prostasomes: their effects on human male reproduction and fertility. *Hum. Reprod. Update.* **12:** 283 – 292.
7. Campos-Silva P, Costa WS, Sampaio FJB, Gregorio BM. Prenatal and/or postnatal high-fat diet alters testicular parameters in adult Wistar Albino rats. Histol Histopathol. (2017) 33:407–16.
8. DeAngelis, A. M.; Roy-O'Reilly, M.; Rodriguez, A. (2014). Genetic Alterations Affecting Cholesterol Metabolism and Human Fertility. Biology of Reproduction, 91(5), 117–117.
9. Dimopoulou C, Goulis DG, Corona G, Maggi M. The complex association between metabolic syndrome and male hypogonadism. Metab Clin Exp.(2018).
10. Eisenberg ML, Kim S, Chen Z, Sundaram R, Schisterman EF, Louis GMB. The relationship between male BMI and waist circumference on semen quality: data from the LIFE study. Hum Reprod. (2015) 30:493–4.
11. Fan Y, Liu Y, Xue K, Gu G, Fan W, Xu Y, et al. Diet-induced obesity in male C57BL/6 mice decreases fertility as a consequence of disrupted bloodtestis barrier. PLoS ONE (2015) 10:e0120775.

12. Fofana M, Travert C, Carreau S and Le Goff D: Evaluation of cholesteryl ester transfer in the seminiferous tubule cells of immature rats in vivo and in vitro. J Reprod Fertil 118: 7983, 2000.

13. Fullston T, McPherson NO, Owens JA, Kang WX, Sandeman LY, Lane M. Paternal obesity induces metabolic and sperm disturbances in male offspring that are exacerbated by their exposure to an "obesogenic" diet. Physiol Rep. (2015) 3:12336.

14. Grondahl, C. 2008. Oocyte maturation. Basic and clinical aspects of in vitro maturation (IVM) with special emphasis of the role of FF-MAS. Dan. Med. Bull. 55: 1 -16 .

15. Jafaroghli M, Abdi-Benemar H, Zamiri MJ, Khalili B, Farshad A, Shadparvar AA. Effects of dietary n-3 fatty acids and vitamin C on semen characteristics, lipid composition of sperm and blood metabolites in fat-tailed Moghani rams. Anim Reprod Sci. (2014) 147:17–24.

16. Jin SK, Yang WX. Factors and pathways involved in capacitation: how are they regulated? Oncotarget (2017) 8:3600–27.

17. Jungwirth A, Giwercman A, Tournaye H, Diemer T, Kopa Z, Dohle G,et al. European Association of Urology Working Group on Male Infertility. European Association of Urology guidelines on Male Infertility: the 2012 update. Eur Urol. (2012) 62:324–32.

18. Kawano, N., K. Yoshida, K. Miyado, and M. Yoshida. 2011. Lipid rafts: keys to sperm maturation, fertilization, and early embryogenesis. *J. Lipids.* **2011:** 264706.

19. Keber R, Rozman D, Horvat S. Sterols in spermatogenesis and sperm maturation. J Lipid Res. 2013 Jan;54(1):20-33.

20. Lobaccaro, Jean-Marc & Gallot, Denis & Lumbroso, Serge & Mouzat, Kevin. (2012). Liver X Receptors and female reproduction: When cholesterol meets fertility!. Journal of endocrinological investigation. 36.

21. Lu JC, Jing J, Yao Q, Fan K, Wang GH, Feng RX, et al. Relationship between Lipids Levels of Serum and Seminal Plasma and Semen Parameters in 631 Chinese Subfertile Men. PLoS ONE (2016) 11:e0146304.

22. Maheshwari A, Stofberg L, Bhattacharya S. Effect of overweight and obesity on assisted reproductive technology--a systematic review.Hum Reprod Update 2007, 13: 433-44.

46. Vignozzi L, Morelli A, Sarchielli E, Comeglio P, Filippi S, Cellai I, et al. Testosterone protects from metabolic syndrome-associated prostate inflammation: an experimental study in rabbit. J Endocrinol. (2012) 212:71–84.
47. Volle DH, Mouzat K, Duggavathi R, Siddeek B, Déchelotte P, Sion B,et al. Multiple roles of the nuclear receptors for oxysterols liver X receptor to maintain male fertility. Mol Endocrinol. (2007) 21:1014–27.
48. Wang F, Chen Z, Ren X, Tian Y, Wang F, Liu C, et al. Hormone-sensitive lipase deficiency alters gene expression and cholesterol content of mouse testis. Reproduction (2017) 153:175–85.
49. Whitfield M, Pollet-Villard X, Levy R, Drevet JR, Saez F. Posttesticular sperm maturation, infertility, and hypercholesterolemia. Asian J Androl. (2015) 17:742–8.
50. Zalata, A., A. Hassan, A. Christophe, F. Comhaire, and T. Mostafa .2010 . Cholesterol and desmosterol in two sperm populations separated on Sil-Select gradient. *Int. J. Androl.* **33:** 528 – 535.

INDEX

E

F

G

H

I

L